Essential Statistics Using SAS® University Edition

Geoff Der · Brian S. Everitt

support.sas.com/bookstore

The correct bibliographic citation for this manual is as follows: Der, Geoff, and Brian S. Everitt. 2015. *Essential Statistics Using SAS® University Edition*. Cary, NC: SAS Institute Inc.

Essential Statistics Using SAS® University Edition

Copyright © 2015, SAS Institute Inc., Cary, NC, USA

ISBN 978-1-62959-843-7 (Hard copy)
ISBN 978-1-62960-094-9 (EPUB)
ISBN 978-1-62960-095-6 (MOBI)
ISBN 978-1-62960-096-3 (PDF)

All Rights Reserved. Produced in the United States of America.

For a hard copy book: No part of this publication may be reproduced, stored in a retrieval system, or transmitted, in any form or by any means, electronic, mechanical, photocopying, or otherwise, without the prior written permission of the publisher, SAS Institute Inc.

For a web download or e-book: Your use of this publication shall be governed by the terms established by the vendor at the time you acquire this publication.

The scanning, uploading, and distribution of this book via the Internet or any other means without the permission of the publisher is illegal and punishable by law. Please purchase only authorized electronic editions and do not participate in or encourage electronic piracy of copyrighted materials. Your support of others' rights is appreciated.

U.S. Government License Rights; Restricted Rights: The Software and its documentation is commercial computer software developed at private expense and is provided with RESTRICTED RIGHTS to the United States Government. Use, duplication, or disclosure of the Software by the United States Government is subject to the license terms of this Agreement pursuant to, as applicable, FAR 12.212, DFAR 227.7202-1(a), DFAR 227.7202-3(a), and DFAR 227.7202-4, and, to the extent required under U.S. federal law, the minimum restricted rights as set out in FAR 52.227-19 (DEC 2007). If FAR 52.227-19 is applicable, this provision serves as notice under clause (c) thereof and no other notice is required to be affixed to the Software or documentation. The Government's rights in Software and documentation shall be only those set forth in this Agreement.

SAS Institute Inc., SAS Campus Drive, Cary, NC 27513-2414

February 2016

SAS® and all other SAS Institute Inc. product or service names are registered trademarks or trademarks of SAS Institute Inc. in the USA and other countries. ® indicates USA registration.

Other brand and product names are trademarks of their respective companies.

SAS software may be provided with certain third-party software, including but not limited to open-source software, which is licensed under its applicable third-party software license agreement. For license information about third-party software distributed with SAS software, refer to **http://support.sas.com/thirdpartylicenses**.

Contents

About This Book

Purpose

This book teaches students of many courses and applied research workers how to analyse data using statistical methods with SAS University Edition.

Is This Book for You?

Students of a variety of courses who need to use statistical methods to complete their coursework will benefit from this book, as well as research workers who are not statisticians but need to apply the correct statistical methods to their data using SAS University Edition.

Prerequisites

This book will be most useful if you are attending or have attended an introductory course of statistics.

What's New in This Edition

SAS University Edition includes the SAS products Base SAS®, SAS/STAT®, SAS/IML®, SAS/ACCESS® Interface to PC Files, and SAS® Studio. Teachers, students, adult learners, and academic researchers can access the SAS University Edition software for noncommercial learning purposes. You can download SAS University Edition for free, directly from SAS. After you download it to your PC, Mac, or Linux workstation, SAS works locally on your machine by using virtualization software and your browser, so no Internet access is required. Or you can get SAS University Edition for free via AWS Marketplace (AWS usage fees may apply). The software runs in the cloud - all you need is a browser and an Internet connection. For more information, visit SAS University Edition at www.sas.com.

SAS University Edition features:

- **An intuitive interface** that lets you interact with the software from your PC, Mac or Linux workstation. Learn more about SAS Studio.

- **A powerful programming language** that's easy to learn, easy to use. Learn more about Base SAS.

- **Comprehensive, reliable tools** that include state-of-the-art statistical methods. Learn more about SAS/STAT.

- **A robust, yet flexible matrix programming language** for more in-depth, specialized analysis and exploration. Learn more about SAS/IML.

- **Several time series forecasting procedures** – TIMEDATA, TIMESERIES, ARIMA, ESM, UCM and TIMEID – from SAS/ETS .

- **Out-of-the-box access to PC file formats** for a simplified approach to accessing data. Learn more about SAS/ACCESS.

Additionally, see the SAS University Edition fact sheet.
http://www.sas.com/content/dam/SAS/en_us/doc/factsheet/sas-university-edition-107140.pdf

Scope of This Book

The use of SAS University Edition for applying a range of statistical techniques widely used in data analysis, and the detailed mathematics behind the statistical methods.

About the Examples

Software Used to Develop the Book's Content

SAS University Edition, version 3.4.

Example Code and Data

The book concentrates almost exclusively on using the built-in tasks so that SAS code is kept to a minimum.

All the example datasets used can be downloaded from the authors' pages at
http://support.sas.com/publishing/authors/der.html

http://support.sas.com/publishing/authors/everitt.html

Look for the cover thumbnail of this book, and select Example Code and Data to download the SAS datasets that are used in this book.

If you are unable to access the datasets through the website, send an e-mail to saspress@sas.com.

SAS University Edition

If you are using SAS University Edition to access data and run your programs, then please check the SAS University Edition page to ensure that the software contains the product or products that you need to run the code: http://support.sas.com/software/products/university-edition/index.html.

Output and Graphics Used in This Book

By default, SAS University Edition uses the HTMLBLUE style, and that has been used for the tabular output in the book. The graphics shown in this book were generated using the Dtree style to ensure optimal print clarity. Output styles are explained on page 14. Some people find the grid lines too prominent with the Dtree style; page 73 shows how to turn these off for some plots.

Exercise Solutions

Exercises are included at the end of each chapter.

Additional Help

Although this book illustrates many analyses regularly performed in businesses across industries, questions specific to your aims and issues may arise. To fully support you, SAS Institute and SAS Press offer you the following help resources:

- For questions about topics covered in this book, contact the author through SAS Press:
  - ○ Send questions by email to saspress@sas.com; include the book title in your correspondence.
  - ○ Submit feedback on the author's page at http://support.sas.com/author_feedback.
- For questions about topics in or beyond the scope of this book, post queries to the relevant SAS Support Communities at https://communities.sas.com/welcome.
- SAS Institute maintains a comprehensive website with up-to-date information. One page that is particularly useful to both the novice and the seasoned SAS user is its Knowledge Base. Search for relevant notes in the "Samples and SAS Notes" section of the Knowledge Base at http://support.sas.com/resources.
- Registered SAS users or their organizations can access SAS Customer Support at http://support.sas.com. Here you can pose specific questions to SAS Customer Support; under *Support*, click *Submit a Problem*. You will need to provide an email address to which replies can be sent, identify your organization, and provide a customer site number or license information. This information can be found in your SAS logs.

Keep in Touch

We look forward to hearing from you. We invite questions, comments, and concerns. If you want to contact us about a specific book, please include the book title in your correspondence.

Contact the Author through SAS Press

- By e-mail: saspress@sas.com
- Via the Web: http://support.sas.com/author_feedback

Purchase SAS Books

For a complete list of books available through SAS, visit sas.com/store/books.

- Phone: 1-800-727-0025
- E-mail: sasbook@sas.com

Subscribe to the SAS Training and Book Report

Receive up-to-date information about SAS training, certification, and publications via email by subscribing to the SAS Training & Book Report monthly eNewsletter. Read the archives and subscribe today at http://support.sas.com/community/newsletters/training!

Publish with SAS

SAS is recruiting authors! Are you interested in writing a book? Visit http://support.sas.com/saspress for more information.

About These Authors

 Geoff Der is a research statistician in the Social and Public Health Sciences Unit at the University of Glasgow, Scotland.

 Brian Everitt is Professor Emeritus at King's College, London.

Geoff and Brian have jointly authored *Basic Statistics Using SAS® Enterprise Guide®: A Primer, Applied Medical Statistics Using SAS®, A Handbook of Statistical Analyses Using SAS®, Third Edition,* and *Statistical Analysis of Medical Data Using SAS®.*

Learn more about these authors by visiting their author pages, where you can download free book excerpts, access example code and data, read the latest reviews, get updates, and more:
http://support.sas.com/der
http://support.sas.com/everitt

Preface

The 1960s were a great decade to be a student. Optimism was in the air, Dylan was declaring that the times were changing, and most students were enjoying the chances to experiment with alternative lifestyles from those of their parents. Most—but not all!—students of statistics were burdened with hours of tedious arithmetic with mechanical calculators, which was needed to get results from the application of one or another statistical technique applied to some particular data set; this absorbed so much of their time that many of them (us) could only be envious observers of the good times going on around them. No wonder that statisticians were often considered rather grumpy and introverted by their colleagues!

In the intervening five decades or so, everything has changed. All of the arithmetic required by a statistical analysis is undertaken with the aid of a computer and a statistical package. This means that there is more energy and more time available to be spent on the interpretation of results and reaching a conclusion about what story the data has to tell. A result of this change is that statisticians have become far happier with their lot, and having one as a friend is no longer something to keep under wraps. The authors of this book now have lots of friends.

Statistics is ubiquitous in all fields of study from archaeology to zoology and from art to zero tolerance. The vast majority of undergraduate and postgraduate students will at some time be faced with undertaking a statistical procedure on their own after perhaps having taken an introductory statistics course. In this book, we demonstrate how to use SAS University Edition to apply a variety of statistical methodology from the simple to the not-so-simple to a range of data sets. The statistical techniques used are only briefly introduced and for details of relevant formulae etc., readers are mainly referred to another text on introductory statistics.

Our main aim here is to show how to apply the appropriate statistical method for answering a particular question about a data set using SAS University Edition and in the correct interpretation of the numerical results obtained. The menu driven interface, SAS Studio, is used for all analyses, and no programming knowledge is assumed or needed.

Exercises are included at the end of most chapters; attempting these exercises will solidify what has been learnt in a chapter. All the data sets are available for downloading from the book's companion Web site at http://support.sas.com/everitt and http://support.sas.com/der.

Geoff Der, Glasgow
Brian Everitt, Dulwich
2015

Chapter 1: Statistics and an Introduction to the SAS University Edition

1.1 Introduction

Statistics is a general intellectual method that applies wherever data, variation, and chance appear. It is a fundamental method because data, variation, and chance are omnipresent in modern life. It is an independent discipline with its own core ideas,, rather than, for example, a branch of mathematics. Statistics offers general, fundamental, and independent ways of thinking (Moore, 1998).

There are, of course, less worthy and less formal statements of what the field of statistics is about;

> *"There are two kinds of statistics, the kind you look up and the kind you make up.*
> *(Rex Stout, Death of a Doxy)*
>
> *Do not put your faith in what statistics say until you have carefully considered what*
> *they do not say. (William W. Watt)"*

Quintessentially, statistics is about solving problems. Data (measurements, observations) relevant to these problems are collected and statistical analyses are used to provide (hopefully) useful answers. But the path from data collection to analysis and interpretation is often not straightforward. Most real-life applications of statistical methodology have one or more nonstandard features. In practice, this means that there are few routine statistical questions, although there are questionable statistical routines. Many statistical pitfalls lie in wait for the unwary. Indeed, statistics is perhaps more open to misuse than most other subjects, particularly by the non-statistician with access to powerful statistical software. Because this book is aimed essentially at users of statistics who might not be expert statisticians and who are using the SAS University Edition, we shall try to make clear in subsequent chapters what is and what is not good statistical practice so that readers can avoid such pitfalls.

Although as mentioned in the Preface, we do not intend this book to be an introduction to statistics *per se*, we will in this chapter cover very briefly some basic concepts that will hopefully be useful to readers in the chapters to come.

1.2 Measurements and Observations

The basic material, the *data* that are the foundation of all scientific investigation no matter in what field, are the *measurements* taken and the *observations* made on the things of interest to the investigator (for example, subjects perhaps in psychology, patients in medicine, artefacts in archaeology, or animals in zoology). These measurements will, of course, *vary* between the subjects, patients, artefacts, and so on, and so are usually all referred to as *variables.* For any defendable conclusions to arise from these measurements, they need to be *objective, precise,* and *reproducible* (Fleiss, 1999). Measurements come in a variety of types and the type of measurement will, in part at least, determine the appropriate method of statistical analysis. It is time to say a little about *scales of measurement.*

1.3 Nominal or Categorical Measurements

Nominal measurements allow subjects, patients, and so on to be classified with respect to some characteristic. Examples of such measurements are categories such as marital status (single, married, divorced), sex (male, female), and blood group (A, O, AB, Other). The properties of a nominal scale are

- The categories are mutually exclusive (an individual can belong to only one category).

- The categories have no logical order—numbers can be assigned to categories but merely as convenient labels.

With nominal scaled measurements, we are essentially confined to counting the number of subjects (or whatever we are dealing with) in each category of the scale.

1.4 Ordinal Scale Measurements

The next level of measurement is the *ordinal scale*. This scale has one additional property over those of a nominal scale—a logical ordering of the categories. With such measurements, numbers assigned to the categories can be used to indicate the amount of a characteristic each variable possesses. A psychiatrist might, for example, grade patients on an anxiety scale as "not anxious," "mildly anxious," "moderately anxious," or "severely anxious" and use the numbers 0, 1, 2, and 3 to label the categories, with lower numbers indicating less anxiety. The psychiatrist cannot infer, however, that the difference in anxiety between patients with scores of 0 and 1 is the same as the difference between patients assigned scores of 2 and 3. The scores on an ordinal scale do, however, allow patients to be *ranked* with respect to the characteristic being assessed.

The following are the properties of an ordinal scale:

- The categories are mutually exclusive.
- The categories have some logical order.
- The categories are scaled according to the amount of a particular characteristic that they indicate.

1.5 Interval Scales

The third level of measurement is the *interval scale*. Such scales possess all the properties of an ordinal scale plus the additional property that equal differences between category levels, on any part of the scale, reflect equal differences in the characteristic being measured. An example of such a scale is temperature on the Celsius (C) or Fahrenheit (F) scale; the difference between temperatures of 80 °F and 90 °F represents the same difference in heat as that between temperatures of 30 °C and 40 °C on the Celsius scale. An important point to make about interval scales is that the zero point is simply another point on the scale; it does *not* represent the starting point of the scale or the total absence of the characteristic being measured. This implies that quoting ratios of such variables is not valid.

The properties of an interval scale are as follows:

- The categories are mutually exclusive.
- The categories have a logical order.
- The categories are scaled according to the amount of the characteristic they indicate.

- Equal differences in the characteristic are represented by equal differences in the numbers assigned to the categories.
- The zero point is completely arbitrary.

1.6 Ratio Scales

The final level of measurement is the *ratio scale*. This type of scale has one further property in addition to those listed for interval scales: it possesses a true zero point that represents the absence of the characteristic being measured. Consequently, statements can be made about both the differences on the scale and the ratio of points on the scale. An example is weight, where not only is the difference between 100 kg and 50 kg the same as between 75 kg and 25 kg, but an object weighing 100 kg can be said to be twice as heavy as one weighing 50 kg. Celsius and Fahrenheit temperatures are not ratio scales, so, for example, a weather reporter who says that today with a temperature of 30 °C is twice as hot as the corresponding day last week when the temperature was 15 °C is wrong. But temperature measured on the Kelvin scale, which does have a true zero point (absolute zero or −273 °C) is a ratio scale, so converting the two temperatures from Celsius to Kelvin by simply adding 273 to each to give 288 K and 303 K lets us say, correctly, that a day with a temperature of 30 °C is 303/288=1.05 times as hot as a day with a temperature of 15 °C.

The properties of a ratio scale are

- The categories are mutually exclusive.
- The data categories have a logical order.
- The categories are scaled according to the amount of the characteristic they possess.
- Equal differences in the characteristic being measured are represented by equal differences in the numbers assigned to the categories.
- The zero point represents an absence of the characteristic being measured.

An awareness of the different types of measurement that might be encountered when collecting data is important because the appropriate method of statistical analysis to use can often depend on the type of measurement involved, a point that we shall consider where necessary in the subsequent chapters.

A further classification of variable types is into *response* or *dependent* variables and *explanatory* variables; a variety of statistical techniques are used to investigate the effects of the latter on a response variable of interest (for example, sex and age on IQ). We shall consider such techniques in Chapters 4 and 6 (linear regression) and in Chapter 7 (logistic regression).

1.7 Populations and Samples

Most statistical methods seek to help the investigator draw conclusions (*inferences*) about a *population* of interest based on a *sample* of observations from that population. For example, we might be interested the height of men of age 70 born in the county of Essex in the United Kingdom in 1944; this is our population. And we might then begin our investigation by measuring the heights of 100 men from this population; these 100 men constitute our sample. (The sample is usually considered to have been taken at *random*; that is, each member of the population has the same chance of being included in the sample, but this is a detail that we will not elaborate on further.) We can now use the sample values of height to say something about the average height in the population. This could be formulated in terms of a hypothesis about the population average, that it is 6 feet, and a suitable *significance test* applied to see whether the sample values of height suggested evidence against the hypothesis. Or we might use the sample values to *estimate* the population average or, better, to produce a range of likely values for the population average, a range know as a *confidence interval*. Neither a significance test nor a confidence interval can give you certainty because, of course, they are based on only a sample of the values from the population; however, the larger the size of the sample, the more credence you can give to the results. (This is probably a good time to remind ourselves of that old Chinese proverb: "*To be uncertain is uncomfortable, but to be certain is to be ridiculous.*")

A variety of significance tests and confidence intervals will appear throughout subsequent chapters.

Statistical methods often require extensive calculations and so require some friendly software for them to be applied to data. And many statistical methods are quintessentially graphical and so require the software to have the ability to construct a variety of plots. Cue the SAS University Edition.

1.8 SAS University Edition

SAS University Edition is a powerful statistical package provided free of cost by SAS Institute to universities.

The user interface for SAS University Edition (known as SAS Studio) is a browser interface that you access using the web browser of your computer. The common browsers —Internet Explorer, Safari, Firefox, and Chrome—are all supported. An advantage of a browser interface is that the computer doing the processing could be anywhere on the web. Usually, it will be the same PC as the browser is running on but could equally well be a remote computer or a cloud computing service, such as Amazon Web Services. In the latter case, the browser could be running on a mobile device such as a tablet.

Installing SAS University Edition on a Windows PC is covered in a document entitled "SASUniversityEditionInstallGuideWindows.pdf".

On a PC, SAS University Edition runs under a virtual machine and requires virtualisation software to be installed. For Windows, there is a choice of Oracle VM VirtualBox and VMware Player. SAS currently recommends VirtualBox.

An important part of the installation is the creation of shared folders. We recommend using `c:\SASUniversityEdition\myfolders` as the path and `myfolders` as the name. Take care to match the case exactly.

1.8.1 Starting SAS University Edition

To start SAS University Edition under Windows:

1. Start VirtualBox or VMware Player.
2. Start (or play) the SAS University Edition virtual machine. It takes a little time to start and then shows an address to enter in your browser. For VirtualBox, this is likely to be http://localhost:10080; for VMware Player, it will be an IP address such as http://172.16.49.136.
3. Note this address.
4. Minimize VirtualBox or VMware Player (do not close it).
5. Start your browser and enter the address above in the address box (the initial http:// might be optional).
6. At the welcome screen, click **Start SAS Studio**.

1.8.2 The SAS Studio Interface

The SAS Studio interface is shown in Display 1.1 and consists of a navigation pane on the left and a work area on the right. In the view shown in Display 1.1, the work area is itself split into two parts: the settings pane and the results pane. Display 1.1 is based on an example in Chapter 4, Section 4.2.3.

Display 1.1: The SAS Studio Interface

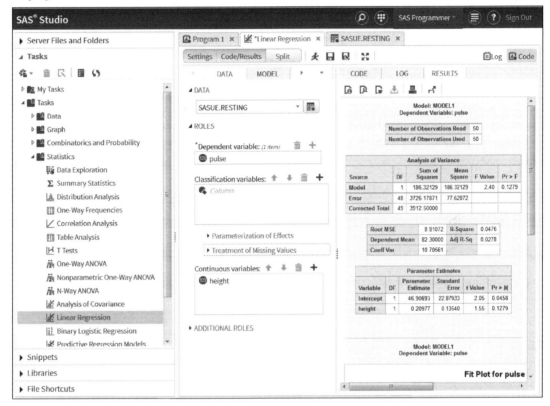

1.8.3 The Navigation Pane

The navigation pane has the following areas: Server Files and Folders, Tasks, Snippets, Libraries, and File Shortcuts. Clicking on the title of an area expands it to show its contents. In Display 1.1, the Tasks area has been expanded. The Tasks area contains menus for data manipulation, graphics, and statistical analyses, and these form the main focus of this book. In Display 1.1, the **Statistics** tasks are shown within the Tasks area and the **Linear Regression** task is selected. The Libraries area allows the contents of SAS data sets to be viewed. The Snippets area contains saved pieces of SAS syntax (also known as *code*) that can be edited. Code snippets are typically used to do things that are not available as tasks. Usually, these require an understanding of SAS syntax, which is beyond the limit of this book, although interested readers can consult one of our other books. The Server File and Folders area is for manipulation of files in a way that will be familiar to most readers. Perhaps its most important use is to verify that files are visible to SAS University Edition in the **myfolders** folder. File Shortcuts are the SAS equivalent of Windows shortcuts.

1.8.4 The Work Area

The work area is where the settings for a task are entered and where the results are shown when the task has run. In Display 1.1, the work area covers the two-thirds of the screen on the right. The middle section, the settings pane, shows some of the settings for the linear regression task and the right-hand section, the results pane, shows the results of the analysis.

The work area is also where the contents of SAS data sets can be displayed. At the top of the work area in Display 1.1 are three tabs: **Program 1**, **Linear Regression**, and **sasue.resting.** The **Program 1** tab is there by default when the software is started. The tab for **sasue.resting** shows that the sasue.resting data set—the one being analysed—has been opened, but its contents are hidden behind those for the linear regression analysis. Given that the data set has already been opened, to view the contents, simply click on the tab, which brings it to the front.

Below the three tabs at the top of the work area is a second line with three more tabs and some icons. The three tabs are labelled **Settings**, **Code/Results**, and **Split**, and they determine what is shown in the work area. If **Settings** is selected, the whole work area is reserved for the task settings. If **Code/Results** is selected, the work area is reserved for code (SAS syntax), and/or the task results. The **Split** tab, selected in Display 1.1, splits the work area into two panes, one on the left for task settings and the other on the right for results. The righthand pane can also show the code and/or log by clicking on the icons on the right-hand end of the line containing the three tabs.

We recommend using the split view for most purposes.

Sometimes more space is needed for the task settings than is available in the standard split view, although this does depend on the size of the screen. One option here would be to temporarily switch to the settings view, but a better option is to maximize the work area by clicking the **Maximize View** button (⚎) or pressing Alt-F11, which enlarges the work area to cover the whole screen. Clicking the same button returns the work area to its previous size.

The other important icon on this line is the **Run** button (🏃); clicking this button runs the task to produce the results. Pressing the F3 function key has the same effect.

Further details of the SAS Studio interface can best be illustrated by an example task. As in Display 1.1, we will use the example from Chapter 4, Section 4.2.3.

1.8.5 Tasks and Task Settings

Opening a Task

The first step in using a task is to open the task in the work area. In this book, we will use tasks for statistical analysis, graphics, and data manipulation. Those tasks are found in the navigation pane under **Statistics**, **Graph**, or **Data**, respectively. In Display 1.1, the Statistics area of the navigation pane has been expanded and the Linear Regression task is highlighted. To expand an area of the navigation pane, click on the triangle (▷) on its left; to collapse it, click on the triangle again (which now looks like this: ◢). To open a task in the work area, either double-click on it, right-click it and select **Open**, or drag it across to the work area.

Entering the Task Settings

Having opened a task, the next step is to enter the task settings. The settings that are shown and that need to be entered vary according to the task. Display 1.2 shows the same example task as Display 1.1 as we begin to enter the settings, but in this case the work area has been expanded and the settings pane enlarged (by dragging the pane border). We can see that the Linear Regression task has tabs for **Data**, **Model**, **Options**, **Selection**, **Output**, and **Information**.

Display 1.2: The SAS Studio Interface with the Work Area Maximized

1.8.6 The Data Tab

The most important is the **Data** tab; nearly all tasks have a **Data** tab and many will only require settings within the **Data** tab. Within it, the first setting specifies the data set that is to be analysed. In Display 1.2, we can see that the data set named in the box is **sasue.resting**. To select a data set for analysis, click on the icon by the box (⊞); a popup window then shows the available libraries, and a data set can be selected from one of these by double-clicking on it.

The **Data** tab is also the place where the variables to be analysed are specified, and these will be assigned roles depending on the task. In a regression type task, there will usually be one dependent variable and one or more predictor variables. The predictor variables might be subdivided into classification variables or continuous variables. Classification variables are nominal scale variables and continuous variables are interval or ratio scale variables. The treatment of ordinal variables depends on the analysis. In Display 1.2, Pulse is the dependent variable and Height a continuous predictor variable.

To assign a variable to a particular role, click the **Add** button (✚) and select it from the popup list of variables in the data set. To remove a variable, select it and click the **Delete** button (🛍).

As settings are entered on the right of the work area, SAS code is generated on the left. In Display 1.2, no code has been generated. Instead, there are comments to say why not. The message is not fully visible in Display 1.2, but it reads: `Add one or more effects to the candidate model on the MODEL tab`. It is also worth noting that the **Run** button (🏃) is greyed out at this point, indicating that the analysis is not ready to be run.

1.8.7 The Model Tab

Display 1.3 shows the next step in entering the settings for this example. The **Model** tab has been selected and the settings pane enlarged further by dragging the dividing bar. The predictor variable, Height, has been added to the **Variables** box on the left automatically by virtue of being assigned that role under the **Data** tab. It has been added to the **Model effects** by selecting it and clicking the **Add** button. There are now sufficient settings entered to generate SAS code, which is shown on the right and the **Run** button is no longer greyed out, indicating that the analysis is ready to run.

Display 1.3: Entering the Model Effect Settings

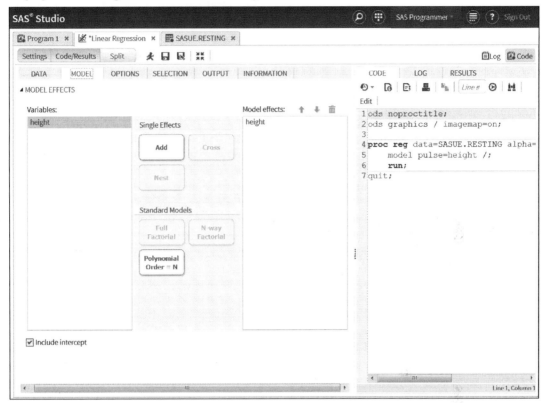

1.8.8 The Options Tab

Even though the task is now ready to run, we might want to alter some of the remaining settings, which have been set by default. In the example given in Chapter 4, we do make some changes. Display 1.4 shows these: the **Options** tab has been selected and in the **Plots** section, the **Fit plot for a single continuous variable** has been selected. **Diagnostic plots, Residuals for each explanatory variable**, and **Observed values by predicted values** had been selected by default, but we deselected them. Most tasks have an **Options** tab, which typically contains sections for plots and statistics. These allow you to select additional plots and statistics as well as to deselect the default ones when they are not required.

Display 1.4: Entering Additional Task Options

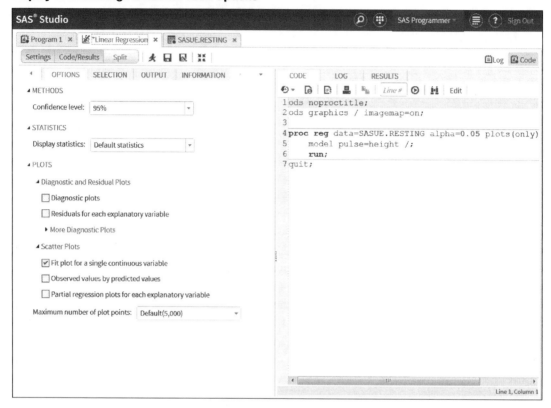

1.8.9 The Output Tab

Some tasks have the ability to produce new data sets as part of their results. These are usually a copy of the input data set with new variables added for such things as predicted values and residuals. The **Output** tab controls whether such a data set is produced and allows it to be named and the additional variables that it contains specified.

1.8.10 The Information Tab

The **Information** tab gives a brief description of the task and the underlying SAS procedures that it uses.

Other Tabs

The **Data**, **Options**, and **Information** tabs are the most common, followed by the **Output** tab. The Linear Regression task also has a **Selection** tab. This is specific to certain regression tasks that have the facility for selecting a subset of predictors from a larger number.

1.8.11 Abbreviations of Task Settings Used in This Book

When describing the settings needed for a particular analysis, we use an abbreviated form of instructions. The example below is the abbreviated form of the instructions for the example described in more detail above:

1. Open **Tasks ▶ Statistics ▶ Linear Regression**.
2. Under **Data ▶ Data**, enter **sasue.resting**.
3. Under **Data ▶ Roles** add **pulse** as the **Dependent variable** and **height** to the **Continuous variables** box.
4. Under **Model ▶ Model Effects**, select **height** and click the **Add** button.
5. Under **Options ▶ Plots ▶ Scatter plots**, select **Fit plot for a single continuous variable** and deselect the other plots.
6. Click the **Run** button.

The first instruction is to open the required task. In this case, it is the Linear Regression task and is to be found in the Statistics group of tasks. Tasks are located in the navigation pane so that pane must be visible. Occasionally, a set of instructions will begin with an instruction to reopen the task. This will be the case where a task has been run and we want to rerun it after changing some settings. In this case, all that is needed is to click on the task's tab in the work area.

The second instruction is shorthand for "enter **sasue.resting** in the Data area under the **Data** tab". Similarly, the third describes the settings for the variable roles in the **Data** tab and the fourth and fifth describe settings in the **Model** and **Options** tabs.

The final instruction is to run the task (that is, to click on the Run button or press F3).

1.8.12 The Results Pane

In Display 1.1, the pane on the right shows the results of running the task. The scroll bars on the edges of the pane indicate that the results will not fully fit within the pane at that size. Rather than using the scroll bars, a better option for viewing results is to have the work area maximised via the **Maximize View** button (⬚) mentioned above. Another option is to display the results in a new browser tab by clicking the right-hand button (↗).

Each time a task is run, the contents of the results pane are replaced. Displaying the results in separate browser tabs is helpful when comparing two or more sets of results. Results can also be viewed, or saved to a file, in any of three formats, HTML (⬚), PDF (⬚), or RTF (⬚), by clicking the appropriate button. RTF– rich text format – is a format designed for importing into word processing programmes. Settings for controlling the format and appearance of results are described below. Results can also be printed (⬚).

In Display 1.1, the results pane also has tabs for **Code** and **Log**. These tabs can be toggled on and off with the **Log** (⊟) and **Code** (⊟) buttons. For the most part, we will be relying on the tasks and task settings to generate the necessary code and can therefore ignore the Code tab. However, as seen, the code pane can provide useful information if we have not entered all the settings necessary for a task. Occasionally, we will edit the code generated by the task in order to access options or settings that are not available in the task settings pane. The **Log** tab is useful for showing whether any errors or warnings were issued when the task was run. Errors are relatively rare when using the tasks, but it is worth checking the log when in doubt.

1.8.13 Options and Preferences

Various aspects of the interface can be configured via the application options button (≡) on the title bar. Clicking this button opens a drop-down menu with sections for **Edit Autoexec File, View, Preferences, Tools**, and **Reset SAS Session**. The autoexec file contains SAS code that is run every time the SAS University Edition is started and can safely be ignored by most users. The View section allows areas of the navigation pane to be hidden. For example, once the data files have been set up for this book, the only areas of the navigation pane that will be needed are the tasks and libraries areas, so Folders, Snippets, and File Shortcuts could all be hidden.

Preferences

The **Preferences** sub-menu opens a popup window with four groups of options: **General**, **Editor**, **Results**, and **Tasks**. The Results section with its default settings is shown in Display 1.5. There are two main points to note: PDF and RTF format output can be turned off and the output style can be chosen separately for each format of output.

To turn off PDF or RTF output, deselect the **Produce PDF output** box or the **Produce RTF output** box. The corresponding button (⧉ or ⧉) in the Results pane will then be greyed out. The HTML output cannot be turned off because the software's interface is a browser interface, so HTML format is its native format. In practice, most users will want to leave RTF output on as this is the most convenient way to transfer results to a word processor document.

Under **Preferences ▶ Tasks**, one setting we recommend is to deselect **Generate header comments for task code.**

Output Styles

The output style determines the appearance of the output (that is, the fonts, colours, and layout of the output). Clicking the drop-down button (▼) shows the styles available. The choice of output style is largely a matter of personal preference. The **Htmlblue** style, which is the default for the Results pane, is a good general choice, as is the **Statistical** style. If black and white output is required, one of the **Journal** styles might be suitable.

As Display 1.5 shows, HTML, PDF, and RTF each have a different default style. However, there are practical advantages to having them set to the same style if, for example, results are being copied to a word processor document.

Display 1.5: The Results Section of the Preferences Menu

1.8.14 Setting Up the Data Used in This Book

Download the online material for this book, which is available at either of our author pages on the support.sas.com web site:

https://support.sas.com/publishing/authors/der.html

http://support.sas.com/publishing/authors/everitt.html

The ZIP file will contain a folder named sasdata with numerous files within it having the extension .sas7bdat. These are the SAS data sets used in this book.

The instructions for installing SAS University Edition included setting up a shared folder. The recommended path for this was C:\SASUniversityEdition\myfolders and the name for the shared folder myfolders. (If you did not create a shared folder at installation, you will need to do it now.)

1. Start SAS University Edition.

2. Under **Server files and folders ▶ My Folders**, click the **New** button (▦), select **Folder** from the drop-down menu, type **sasdata** for the name, and click **Save**.

3. Select the newly created **sasdata** folder and click the **Upload Data** (⬆) button.

4. Click **Choose files**, and navigate to the place where the downloaded online materials were saved.

5. Select all sasdata files (that is, those with the extension .sas7bdat), and click **Upload**.

   (On a PC, it is also possible to simply copy the downloaded **sasdata** folder to `C:\SASUniversityEdition\myfolders`.)

Having copied or uploaded the SAS data files, we now need to assign them to a library.

1. In the navigation area under **Libraries**, click the **New Library** button (▤).

2. In the box for the name, enter **sasue**.

3. For the path, enter **/folders/myfolders/sasdata**.

4. Select **Re-create this library at start-up.**

Chapter 2: Data Description and Simple Inference

2.1 Introduction

In this chapter, we will describe how to get informative numerical summaries of data and graphs that allow us to assess various properties of the data. In addition, we will show how to test whether different populations have the same mean value for some variable of interest. The statistical topics covered are:

1. Summary statistics such as means and variances
2. Graphs such as histograms and box plots
3. Student's *t*-test

2.2 Summary Statistics and Graphical Representations of Data

Shortly after metric units of length were officially introduced in Australia in the 1970s, each of a group of 44 students was asked to guess, to the nearest metre, the width of the lecture hall in which they were sitting. Another group of 69 students in the same room was asked to guess the width in feet, to the nearest foot. The measured width of the room was 13.1 metres (43.0 feet). The data were collected by Professor T. Lewis, and are given here in Table 2.1, which is taken from Hand et al. (1994). Of primary interest here is whether the guesses made in metres differ from the guesses made in feet, and which set of guesses gives the most accurate assessment of the true width of the room (accuracy in this context implies guesses that are closer to the measured width of the room).

Table 2.1: Room Width Estimates

Guesses in Metres

8	9	10	10	10	10	10	10	11	11	11
11	12	12	13	13	13	14	14	14	15	15
15	15	15	15	15	15	16	16	16	17	17
17	17	18	18	20	22	25	27	35	38	40

Guesses in Feet

24	25	27	30	30	30	30	30	30	32	32
33	34	34	34	35	35	36	36	36	37	37
40	40	40	40	40	40	40	40	40	41	41
42	42	42	43	43	44	44	44	44	45	45
45	45	45	45	46	46	47	48	48	50	50
50	51	54	54	54	55	55	60	60	63	70
75	80	94								

2.2.1 Initial Analysis of Room Width Guesses Using Simple Summary Statistics and Graphics

How should we begin our investigation of the room width guesses data that are given in Table 2.1? As with most data sets, the initial data analysis steps should involve the calculation of simple summary statistics, such as means and variances, and graphs and diagrams that convey clearly the general characteristics of the data and perhaps enable any unusual observations or patterns in the data to be detected. The data set **sasue.widths** contains two variables: units and guess. First, we will convert the guesses made in metres into feet by multiplying them by 3.28 and then we will calculate the means and standard deviations of the metre and feet estimates.

To create a new variable with all estimates in feet, we will use a short program. A program tab is opened under **Folders** by clicking on the **New** button ( ) and selecting **SAS program** from the drop-down menu or by pressing F4. In the resulting program window, type and then run the following code;

```
data work.widths;
 set sasue.widths;
 if units='metres' then feet=guess*3.28;
 else feet=guess;
run;
```

This creates a new version of the data set, **work.widths**, with a third variable, feet, which contains the guesses in feet. Data sets in the Work library are temporary in the sense that they are deleted when the session is finished.

Summary statistics can be produced using the SAS University Edition task of that name:

1. Open **Tasks ▶ Statistics ▶ Summary Statistics.**
2. Under **Data ▶ Data**, select **work.widths.**
3. Under **Data ▶ Roles**, select **feet** as the **analysis variable** and **units** as a **Classification variable.**
4. Click **Run.**

The edited results are shown in Output 2.1.

Output 2.1: Summary Statistics for Room Width Guesses Data

		Analysis Variable : feet				
units	N Obs	Mean	Std Dev	Minimum	Maximum	N
feet	69	43.6956522	12.4974166	24.0000000	94.0000000	69
metres	44	52.5545455	23.4344427	26.2400000	131.2000000	44

What do the summary statistics tell us about the two sets of guesses? It appears that the guesses made in feet are closer to the measured room width and less variable than the guesses made in metres, suggesting that the guesses made in the more familiar units, feet, are more accurate than those made in the recently introduced units, meters. But often such apparent differences in means and in variation can be traced to the effect of one or two unusual observations that statisticians like to call *outliers*. Such observations can usually be uncovered by some simple graphics, and here we shall construct *box plots* of the two sets of guesses after converting the guesses made in metres to feet.

A box plot is a graphical display useful for highlighting important distributional features of a continuous measurement. The diagram is based on what is known as the *five-number summary* of a data set, the numbers in question being the minimum, the lower quartile, the median, the upper quartile, and the maximum. The box plot is constructed by first drawing a box with ends at the lower and upper quartiles of the data; next, a horizontal line (or some other feature) is used to indicate the position of the median within the box and then lines are drawn from each end of the box to points defined by the upper quartile plus 1.5 times the *interquartile range* (the difference between the upper and lower quartiles) and the lower quartile minus 1.5 times the interquartile range. Any observations outside these limits are represented individually by some means in the finished graphic, and such observations are likely candidates to be labelled outliers. The resulting diagram schematically represents the body of the data minus the extreme observations and is particularly useful for comparing the distributional features of a measurement made in different groups.

The Summary Statistics task also produces box plots (under **options ▶ Plots ▶ Comparative box plot**), but this does not display outliers in the way described above. Instead, we will use the Box Plot task:

1. Open **Tasks ▶ Graph ▶ Box Plot**.
2. Under **Data ▶ Data**, select **work.widths**.
3. Under **Data ▶ Roles**, add **feet** as the **Analysis variable** and **units** as the **Category variable**.
4. Click **Run**.

The resulting plots are shown in Figure 2.1; they indicate that both sets of guesses contain a number of possible outliers and also that the guesses made in metres are *skewed* (have a longer tail) and are more variable than the guesses made in feet. We shall return to these findings in the next subsection.

Figure 2.1: Box Plots of Room Width Guesses Made in Feet and in Metres (After Conversion to Feet)

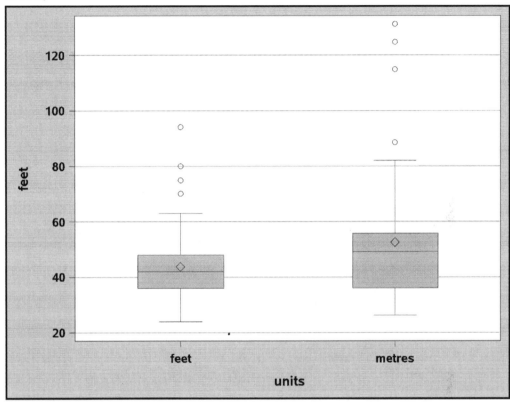

Box plots are our favourite graphic for comparing the distributional properties of a measurement made in different groups, and they are available as optional plots in several of the statistical tasks. Another graphic for displaying distributions is the *histogram*. In a histogram, the range of values is divided into small intervals and the number of observations within each interval is represented by the area of a rectangle centred on the interval; if intervals are all equal, then the heights of the rectangles are proportional to the observed frequencies.

The Histogram task within **Tasks ▶ Graph** can be used to produce a single histogram. For a comparative histogram, we will use the Distribution Analysis task within **Tasks ▶ Statistics**:

1. Open **Tasks ▶ Statistics ▶ Distribution Analysis**.
2. Under **Data ▶ Data**, select **work.widths**.
3. Under **Data ▶ Roles**, add **feet** as the **Analysis variable**.
4. Under **Options ▶ Exploring data ▶ Histogram ▶ Classification Variables**, add **units**.
5. Click **Run**.

The resulting plots are shown in Figure 2.2; they show clearly the greater skewness in the guesses made in metres.

Figure 2.2: Histograms for Room Width Guesses Data

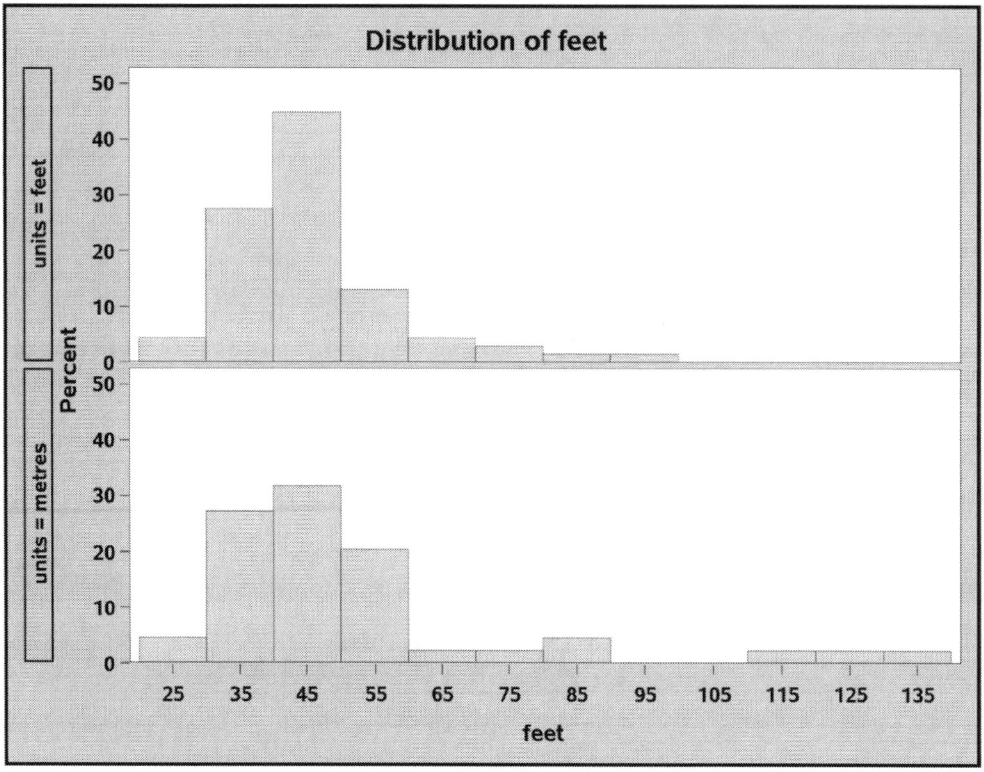

2.3 Testing Hypotheses and Student's *t*-Test

From the summary statistics and graphics produced in the previous subsection, we already know quite a bit about how the guesses of room width made in feet differ from the guesses made in metres. The guesses made in feet appear to be concentrated around the measured room width of 43.0 feet, whereas the guesses made in meters suggest overestimation of the width of the room. In some circumstances, we might simply stop here and try to find an explanation of the apparent difference between the two types of guesses (and many statisticians would be sympathetic to this approach!). But, in general, the investigation of the data will need to go further and use more formal statistical methods to try to confirm our very strong hunch that guesses of room width made in metres differ from guesses made in feet.

The area of statistics that we need to move into is that of *statistical inference*, the process of drawing conclusions about a *population* on the basis of measurements or observations made on a *sample* of observations from the population, a process that is central to statistics. More specifically, inference is about testing hypotheses of interest about some population value on the basis of the sample values, and involves what is known as *significance tests*. For the room width guesses data in Table 2.1, for example, there are three hypotheses that we might wish to test:

In the population of guesses made in metres, the mean is the same as the true room width (namely, 13.1 metres). Formally, we might write this hypothesis as

$$H_0 : \mu_m = 13.1$$

where H_0 denotes the *null hypothesis.*

In the population of guesses made in feet, the mean is the same as the true room width (namely, 43.0 feet), that is

$$H_0 : \mu_f = 43.0$$

After the conversion of metres into feet, the population means of both types of guess are equal, or in formal terms

$$H_0 : \mu_m \text{x} 3.28 = \mu_f$$

(It might be imagined that a conclusion about the last of these three hypotheses would be implied from the results found for the first two, but as we shall see later this is *not* the case.)

2.3.1 Applying Student's *t*-Test to the Guesses of Room Width

Testing hypotheses about population means requires what is known as the *Student's t-test*. The test is described in detail in Altman (1991) but in essence involves the calculation of a *test statistic* from sample means and standard deviations, the distribution of which is known if the null hypothesis is true and certain assumptions are met. From the known distribution of the test statistic, a *p-value* can be found.

The *p*-value is probably the most ubiquitous statistical index found in the applied sciences literature and is particularly widely used in biomedical and psychological research. So just what is the *p*-value? Well, the *p*-value is the probability of obtaining the observed data (or data that represent a more extreme departure from the null hypothesis) if the null hypothesis is true. It was first proposed as part of a quasi-formal method of inference by a famous statistician, Ronald Aylmer Fisher, in his influential 1925 book, *Statistical Methods for Research Workers*. For Fisher, the *p*-value represented an attempt to provide a relatively informal measure of evidence against the null hypothesis; the smaller the *p*-value, the greater the evidence that the null hypothesis is incorrect.

But, sadly, Fisher's informal approach to interpreting the *p*-value has long ago been abandoned in favour of a simple division of results into significant and non-significant on the basis of comparing the *p*-value with some largely arbitrary threshold value such as 0.05. The implication of this division is that there can always be a simple yes (significant) or no (non-significant) answer as the fundamental result from a study; this is clearly false, and used in this way hypothesis testing is of limited value.

In fact, overemphasis on hypothesis testing and the use of *p*-values to dichotomise significant or non-significant results has distracted from other more useful approaches to interpreting study results, in particular the use of *confidence intervals*. Such intervals are a far more useful alternative to *p*-values for presenting results in relation to a statistical null hypothesis and give a range of values for a quantity of interest that includes the population value of the quantity with some specified probability. (Confidence intervals are described in detail in Altman, 1991.) In essence, the significance test and associated *p*-value relate to what the population quantity of interest is *not*; the confidence interval gives a plausible range for what the quantity *is*.

So, after this rather lengthy digression, let's apply the relevant Student's *t*-tests to the three hypotheses that we are interested in assessing on the room width data. The first two hypotheses require the application of the single sample *t*-test, referred to in the software as a one-sample *t*-test, separately to each set of guesses. We start by splitting the widths data set into two parts according to whether the guesses were made in metres or feet:

1. Open **Tasks ▶ Data ▶ Filter Data**.
2. Under **Data ▶ Data**, add **work.widths**.
3. Under **Data ▶ Filter**, assign **units** as variable 1, set the comparison to **Equal**, and type **feet** in the **Value** box.
4. Under **Data ▶ Output Data Se**t, type **feet**.
5. Click **Run**.

The result is a temporary data set, **feet**, in the Work library. Repeat the above, entering **metres** instead of **feet** in steps 3 and 4 (above). Then, to apply the single sample *t*-test to the guesses in feet:

1. Open **Tasks ▶ Statistics ▶ T Tests**.
2. Under **Data ▶ Data**, add **work.feet**.
3. Under **Data ▶ Roles ▶ Analysis variable**, add **guess**. Note that **One-sample test** is the default.
4. Under **Options ▶ Tests ▶ Alternative hypothesis**, enter **43** in the box and deselect **Tests for normality**.
5. Click **Run**.

For the guesses made in metres, repeat the above using **work.metres** in step 2 and **13.1** for the value in step 4.

The results are shown in Output 2.2. Let's now look at these results in some detail. Looking first at the two *p*-values, we see that there is no evidence that the guesses made in feet differ in mean from the true width of the room, 43 feet (the 95% confidence interval here is [40.69, 46.70] feet, which includes the true value of 43 feet). But there is considerable evidence that the guesses made in metres do differ from the true value of 13.1 metres; here, the confidence interval is [13.85, 18.20] and the students appear to systematically overestimate the width of the room when guessing in metres.

Output 2.2: Results of Single Sample *t*-Tests for Room Width Guesses Made in Feet and for Guesses Made in Metres

a) **Guesses Made in Feet**

Mean	95% CL Mean		Std Dev	95% CL Std Dev	
43.6957	40.6934	46.6979	12.4974	10.7044	15.0176

DF	t Value	Pr > |t|
68	0.46	0.6453

b) **Guesses Made in Metres**

Mean	95% CL Mean		Std Dev	95% CL Std Dev	
16.0227	13.8506	18.1949	7.1446	5.9031	9.0525

DF	t Value	Pr > |t|
43	2.71	0.0095

Now it might be thought that our third hypothesis discussed above, namely that the mean of the guesses made in feet and the mean of the guesses made in metres (after conversion to feet) are the same, can be assessed simply from the results given in Output 2.2. Because the population mean of guesses made in feet apparently does not differ from the true width of the lecture room, but the population mean of guesses does differ from the true value, then surely the means of the two groups of guesses differ from each other? Not necessarily, and to assess the equality of means hypothesis correctly, we need to apply an *independent samples t*-test to the data. We again use the *t*-test task:

1. Open **Tasks ▶ Statistics ▶ T Test**.
2. Under **Data ▶ Data**, add **work.widths**.
3. Under **Data ▶ Roles ▶ T test**, select **Two-sample test**.
4. Under **Data ▶ Roles ▶ Analysis variable**, add **feet** as the **Analysis variable** and **units** as the **Groups** variable.
5. Click **Run**.

The results of applying this test are shown in Output 2.3. (The graphical output shown in Figures 2.3 (a) and (b) will be discussed in the next subsection.) Ignoring for the moment the tests for normality, we look first at the *p*-value associated with the *t*-test when equality of variances is assumed; the value is $p=0.0102$, from which we can conclude that there is considerable evidence that the population means of the two types of guesses do indeed differ. The confidence interval for the difference [-15.57,-2.15] indicates that the guesses made in feet have a mean that is between about 15 and 2 feet lower than the guesses made in metres.

Output 2.3: Results of Applying Independent Samples *t*-Test to the Room Width Guesses Data

units = feet

Tests for Normality				
Test		**Statistic**	**p Value**	
Shapiro-Wilk	W	0.886164	Pr < W	<0.0001
Kolmogorov-Smirnov	D	0.15409	Pr > D	<0.0100
Cramer-von Mises	W-Sq	0.313808	Pr > W-Sq	<0.0050
Anderson-Darling	A-Sq	1.911938	Pr > A-Sq	<0.0050

units = metres

Tests for Normality				
Test		**Statistic**	**p Value**	
Shapiro-Wilk	W	0.765685	Pr < W	<0.0001
Kolmogorov-Smirnov	D	0.241055	Pr > D	<0.0100
Cramer-von Mises	W-Sq	0.564252	Pr > W-Sq	<0.0050
Anderson-Darling	A-Sq	3.302602	Pr > A-Sq	<0.0050

units	N	Mean	Std Dev	Std Err	Minimum	Maximum
feet	69	43.6957	12.4974	1.5045	24.0000	94.0000
metres	44	52.5545	23.4344	3.5329	26.2400	131.2
Diff (1-2)		-8.8589	17.5620	3.3881		

units	Method	Mean	95% CL Mean		Std Dev	95% CL Std Dev	
feet		43.6957	40.6934	46.6979	12.4974	10.7044	15.0176
metres		52.5545	45.4298	59.6793	23.4344	19.3621	29.6920
Diff (1-2)	Pooled	-8.8589	-15.5727	-2.1451	17.5620	15.5245	20.2200
Diff (1-2)	Satterthwaite	-8.8589	-16.5431	-1.1747			

Method	Variances	DF	t Value	Pr > |t|
Pooled	Equal	111	-2.61	0.0102
Satterthwaite	Unequal	58.788	-2.31	0.0246

Equality of Variances				
Method	Num DF	Den DF	F Value	Pr > F
Folded F	43	68	3.52	<.0001

Figure 2.3 (a): Histograms of Length Guesses in Feet and in Meters After Conversion to Feet with Fitted Normal Distributions and Kernel Density Function Estimates

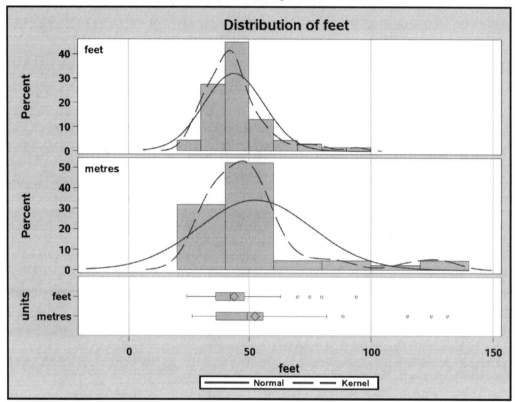

Figure 2.3 (b): Probability Plots of Length Guesses

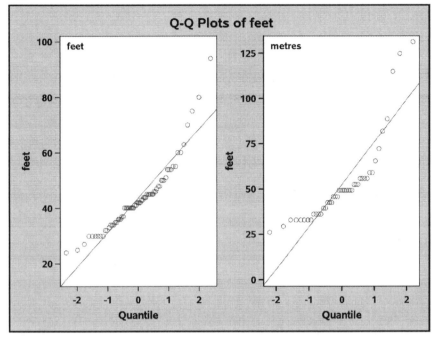

2.3.2 Checking the Assumptions Made When Using Student's *t*-Test and Alternatives to the *t*-Test

Having applied *t*-tests to assess each of the hypotheses of interest and having found the corresponding *p*-values and confidence intervals, it might appear that we have finished the analysis of the room width data. But as yet, we have not looked at the assumptions that underlie the *t*-tests and have not checked whether these assumptions are likely to be valid for the data. First the assumptions:

- The measurements are assumed to be sampled from a *normal distribution*.
- For the independent samples *t*-test, each population is assumed to have the same variance.
- The measurements made are *independent* of each other.

If any (or all) of these assumptions is invalid, then strictly speaking, the *t*-test is also not valid. In practice small departures from the assumptions are unlikely to be of any great importance, but it is still worth trying (at least informally) to check whether the data meet the assumptions. So first let's consider the normality assumption. The results of applying a number of formal tests of normality are shown in Output 2.3. Each of these tests is described in detail in Everitt and Skrondal (2010); they are all highly significant, indicating that the normality assumption is in doubt for length guesses both in feet and in meters. But these formal tests can often be misleading and are rarely used.

Preferable for checking the normality assumption is to use histograms of the data enhanced by fitted normal distributions as shown in Figure 2.3 (a) (the kernel estimate shown in the plot is explained in Der and Everitt, 2013; it is essentially a method of estimating a distribution without assuming any specified distributional form, for example a normal distribution) and what is known as a *probability plot*, which in essence involves a plot of the observed quantiles against theoretical quantiles of the normal distribution (for details, see Everitt and Palmer, 2005). Such plots should have the form of a straight line, that is, they should be *linear*, if the sample does arise from a normal distribution. These plots are shown in Figure 2.3 (b).

Both graphics, but particularly those for the guesses in metres, throw the normality assumption required for the *t*-test to be valid into some doubt. This possible non-normality combined with the evidence that two types of guesses have different variances obtained from both the initial examination of the data and the test for equality of variances (see Altman, 1991) given in Output 2.3 suggests that some caution is needed in interpreting the results from our *t*-tests. Fortunately, the *t*-test is known to be relatively *robust* against departures both from normality and the homogeneity assumption, although it is somewhat difficult to predict how a combination of non-normality, heterogeneity, and outliers will affect the test.

Since the test for equality of variance given in Output 2.3 has an associated *p*-value of <0.001, we should perhaps first consider using a modified version of the *t*-test in which the equality of variance assumption is dropped (see Altman, 1991, for details). The *p*-value of the modified test (*Satterthwaite test*-see Everitt and Skrondal, 2010) is also given in Output 2.3 and, although less significant than the usual form of the *t*-test, still shows evidence for a difference in the population means of the two types of room width guesses.

Here, however, given the existence of outliers in the data and their possible non-normality, we might ask whether an alternative test is available that is both insensitive to the effect of outliers and does not assume normality.

An alternative to Student's *t*-test that does not depend on the assumption of normality is the *Wilcoxon Mann-Whitney test;* this test, which since it is based on the *ranks* of the observations, is also unlikely to be affected greatly by outliers. The Wilcoxon Mann-Whitney test, which is described in detail in Altman (1991), assesses whether the distribution of the measurements in the two groups is the same. This is available in the *t*-test task and is applied as follows: reopen the *t*-tests task or repeat steps 1 to 4 above:

1. Under **Options ▶ Tests**, select **Wilcoxon rank-sum test**.
2. Click **Run**.

The *p*-value for the test is 0.028, confirming the difference in location between the guesses in feet and the guesses in metres.

2.4 The *t*-Test for Paired Data

In a design study for a device to generate electricity from wave power at sea, experiments were carried out on scale models in a wave tank to establish how the choice of mooring method for the system affected the bending stress produced in part of the device. The wave tank could simulate a wide range of sea states (rough, calm, moderate, and so on), and the model system was subjected to the same sample of sea states with each of two mooring methods, one of which was considerably cheaper than the other. The resulting data giving root mean square bending moment in Newton metres are shown in Table 2.2 (these data are taken from Hand et al., 1994). The question of interest is whether bending stress differs for the two mooring methods.

Table 2.2: Wave Energy Device Mooring Data

Sea State	Method I	Method II
1	2.23	1.82
2	2.55	2.42
3	7.99	8.26
4	4.09	3.46
5	9.62	9.77
6	1.59	1.40
7	8.98	8.88
8	0.82	0.87
9	10.83	11.20
10	1.54	1.33
11	10.75	10.32
12	5.79	5.87
13	5.91	6.44
14	5.79	5.87
15	5.50	5.30
16	9.96	9.82
17	1.92	1.69
18	7.38	7.41

2.4.1 Initial Analysis of Wave Energy Data Using Box Plots

For the wave energy data in Table 2.2, we will construct box plots of the bending stresses for each mooring method and here, for reasons that will become apparent in the next subsection, it is also useful to have a look at the box plot of the differences between the pairs of observations made for the same sea state.

The dataset **sasue.waves** contains variables: **state**, **method1**, **method2**, and **difference**.

1. **Graph ▶ Box Plot**
2. Under **Data ▶ Data** enter **sasue.waves**
3. Under **Data ▶ Roles ▶ Analysis variable** add **method1**
4. Click **Run**.

Run the box plot task twice more, once each with method2 and difference as the analysis variable. The results are shown in Figures 2.4 (a), (b) and (c).

Figure 2.4: Box Plots of Root Mean Square Bending Moment (Newton Metres) for Mooring Methods I and II

(a) **Method I**

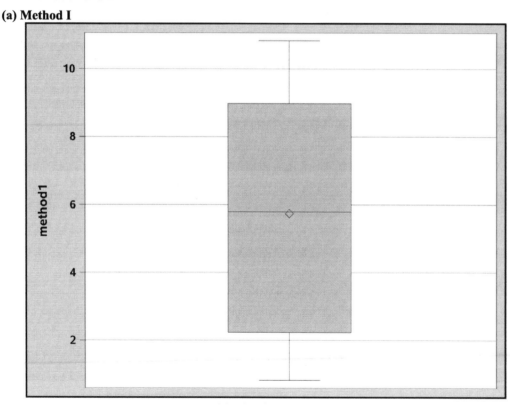

(b) Method II

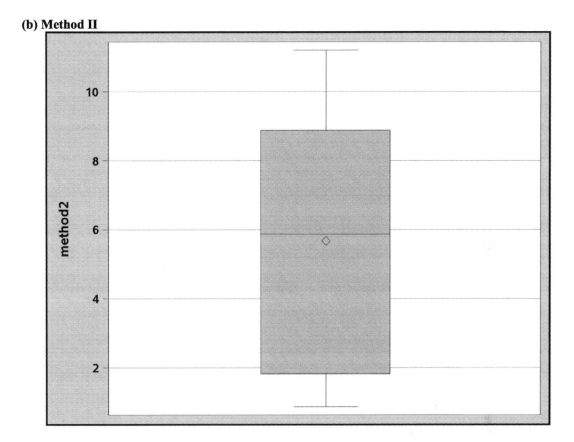

Figure 2.5: Box Plot of Differences of Root Mean Square Bending Moment for the Two Mooring Methods

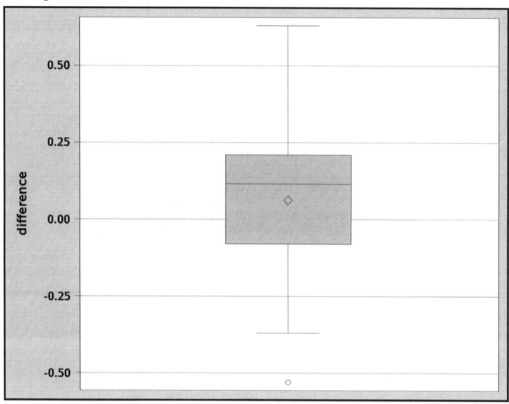

The box plot of differences in Figure 2.5 suggests that there might be one outlying observation that we might wish to check, and a small degree of skewness, although with only 18 observations, drawing any conclusions about the distributional properties of the data is difficult.

2.4.2 Wave Power and Mooring Methods: Do Two Mooring Methods Differ in Bending Stress?

Now we can move on to consider more formally the questions of interest about the wave energy data. Superficially, these data look to be of a very similar format to the room width guesses data, but closer consideration shows that there is a fundamental difference in that the observations are *paired*; that is, the bending stress for each of the two mooring methods is, in each case, based on the same sea state. Consequently, these observations are likely to be correlated rather than independent. So here to test whether there is a difference in the mean bending stress of the two methods of mooring, we use what is called a *paired t-test* (see Altman, 1991, for details); essentially, this test is the same as the one-sample *t*-test used previously for the room width data, but here the null hypothesis is that the population mean of the *differences* of the paired observations is zero. To apply the test:

1. Open **Tasks ▶ Statistics ▶ T Tests**.
2. Under **Data ▶ Data**, add **sasue.waves**.
3. Under **Data ▶ Roles ▶ T Test**, choose **Paired** test.
4. Under **Data ▶ Roles**, select **method1** as the **Group 1 variable** and **method2** as the **Group2 variable**.
5. Under **Options ▶ Tests**, deselect **Tests for normality**.
6. Under **Options ▶ Plots**, choose **selected plots** and select **Histogram and box plot**.
7. **Run**.

The salient numerical results are shown in Output 2.4. The *p*-value associated with the paired *t*-test applied to the wave power data is 0.38. There is no evidence of a difference in mean bending stress for the two mooring methods. The associated graphic in Figure 2.6 indicates that the normality assumption for the differences is justified.

Output 2.4: Results of Applying the Paired t-Test to the Wave Power Data

N	Mean	Std Dev	Std Err	Minimum	Maximum
18	0.0617	0.2901	0.0684	-0.5300	0.6300

Mean	95% CL Mean		Std Dev	95% CL Std Dev	
0.0617	-0.0826	0.2059	0.2901	0.2177	0.4349

DF	t Value	Pr > |t|
17	0.90	0.3797

Figure 2.6: Histogram and Fitted Normal Distribution for the Differences in Bending Stress of the Two Mooring Methods

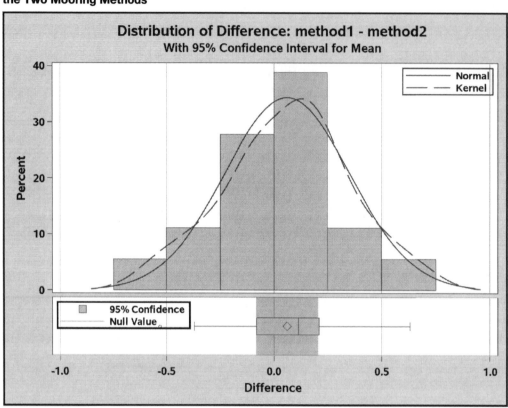

2.4.3 Checking the Assumptions of the Paired *t*-Tests

For the paired *t*-test to be valid, the differences between the paired observations need to be normally distributed. We could use a probability plot to assess the required normality of the differences, but with only 18 observations for the wave data, the plot would not be very useful. Since we cannot satisfactorily assess the normality assumption for these data, we might wish to consider a non-parametric alternative; this would be the *Wilcoxon signed rank test.*

The non-parametric analogue of the paired *t*-test is Wilcoxon's signed rank test (see Altman, 1991, for details). As with the Wilcoxon Mann-Whitney test described above, the signed rank test uses only the ranks of the observations and does not assume normality for the observations; the test is available as an option for the *t*-Tests task:

1. Reopen the T tests task or repeat steps 1 to 6 above.
2. Under **Options ▶ Tests**, select **Sign test** and **Wilcoxon signed rank test**.
3. Click **Run**.

This is also an alternative way of applying a matched-pairs *t*-test, as can be seen from the results in Output 2.5.

Output 2.5: Wilcoxon Signed Rank Test for Wave Energy Mooring Data

Tests for Location: Mu0=0				
Test		**Statistic**	**p Value**	
Student's t	t	0.901934	Pr > |t|	0.3797
Sign	M	1	Pr >= |M|	0.8145
Signed Rank	S	23.5	Pr >= |S|	0.3194

The test gives a *p*-value of 0.319, confirming the result from the paired *t*-test.

2.5 Exercises

Exercise 2.1: Babies Data

The data set, **babies**, gives the recorded birth weights of 50 infants who displayed severe idiopathic respiratory distress syndrome (SIRDS). SIRDS is a serious condition that can result in death and did so in the case of 27 of these children. One of the questions of interest about these data is whether the babies who died differed in birth weight from the babies who survived. Use some suitable graphical techniques to carry out an initial analysis of these data and then find a 95% confidence interval for the difference in mean birth weight for SIRDS babies who die and SIRDS babies who live.

Birth Weights (kg)

Survived							
1.130	1.575	1.680	1.760	1.930	2.015	2.090	2.600
2.700	2.950	3.160	3.400	3.640	2.830	1.410	1.715
1.720	2.040	2.200	2.400	2.550	2.570	3.005	
Died							
1.050	1.175	1.230	1.310	1.500	1.600	1.720	1.750
1.770	2.275	2.500	1.030	1.100	1.185	1.225	1.262
1.295	1.300	1.550	1.820	1.890	1.940	2.200	2.270
2.440	2.560	2.370					

Exercise 2.2: Choles Data

The data in the **choles** data set were collected by the Western Collaborative Group Study carried out in California in 1960-1961. In this study, 3,154 middle-aged men were used to investigate the possible relationship between behaviour pattern and risk of coronary heart disease. The data set contains data from the 38 heaviest men in the study (all weighing at least 225 pounds). Cholesterol measurements (mg/100ml) and behaviour type were recorded; type A behaviour is characterized by urgency, aggression, and ambition, and type B behaviour is relaxed, non-competitive, and less hurried. The question of interest is whether, in heavy middle-aged men, cholesterol level is related to behaviour type. Investigate the question of interest in any way that you feel is appropriate, paying particular attention to assumptions and to any observations that might possibly distort conclusions.

Type A:

233 291 312 250 246 197 268 224 329 239 254 276 234 181 248 252 202 218 325

Type B:

420 185 263 246 224 212 188 250 148 169 226 175 242 153 183 137 202 194 213

Exercise 2.3: Diet Data

The data in **diet** come from a study of the Stillman diet, a diet that consists primarily of protein and animal fats and that restricts carbohydrate intake. In **diet**, triglyceride values (mg/100ml) are given for 16 participants both before beginning the diet and at the end of a period of time following the diet. Here interest is on whether there has been a change in triglyceride level that might be attributed to the diet. Carry out an appropriate hypothesis test to investigate whether there has been a change in triglyceride level, using any graphics that you think might be helpful in interpreting the test.

Subject	Baseline	Final
1	159	194
2	93	122
3	130	158
4	174	154
5	148	93
6	148	90
7	85	101
8	180	99
9	92	183

Subject	Baseline	Final
10	89	82
11	204	100
12	182	104
13	110	72
14	88	108
15	134	110
16	84	81

Chapter 3: Categorical Data

3.1 Introduction

In this chapter, we discuss how to deal with various aspects of the analysis of data containing *categorical variables* (that is, variables that classify the observations in some way). Some examples of categorical variables are gender, marital status, and social class. Numbers might be used as convenient labels for the categories of categorical variables but have no numerical significance. When using categorical variables, we can simply count the number of our sample that fall into each category of a variable or into a combination of the categories of two or more categorical variables. In this chapter, the statistical topics to be covered are:

- Graphical summary of one-way tables, bar charts, and pie charts
- Testing for association of two categorical variables--chi-squared tests for independence
- Testing for association of two categorical variables when some observed counts are small--Fisher's exact test
- Testing for equal probability of an event in matched-pairs data--McNemar's test

3.2 Graphing and Analysing Frequencies: Horse Race Winners

The data shown in Table 3.1 show the starting stall of the winners in 144 horse races held in the USA; all 144 races took place on a circular track and all relate to eight horse races. Starting stall 1 is closest to the rail on the inside of the track. Interest here lies in assessing how the chances of a horse winning a race are affected by its position in the starting line up.

Table 3.1: Horse Racing Data after Classification

Starting stall	2	3	4	5	6	7	8
Number of winners	19	18	25	17	10	15	11

3.2.1 Looking at Horse Race Winners Using Some Simple Graphics: Bar Charts and Pie Charts

The horse racing data are in a SAS data set called **racestalls**, which contains a single variable giving the stall number for each of the 144 winners. To open the data set;

1. Select the **Libraries** section in the navigator pane.
2. Expand **My Libraries** and **SASUE**, scroll down, and double-click on **racestalls**.

There are 144 rows but only one column that is called **stall**.

We can now reproduce Table 3.1 showing the number of winners from each of the eight starting stalls and the corresponding percentages:

1. Open **Tasks ▶ Statistics ▶ One Way Frequencies**.
2. Under **Data ▶ Data**, add **sasue.racestalls**.
3. Under **Data ▶ Roles ▶ Analysis variables**, add **stall**.
4. Click **Run**.

The results begin with the frequencies and percentages shown in Output 3.1. We see that the percentage of winning horses from each stall differs considerably suggesting that the stall does play a part in determining which horse will win.

Output 3.1: Horse Racing Data

Stall	Frequency	Percent	Cumulative Frequency	Cumulative Percent
1	29	20.14	29	20.14
2	19	13.19	48	33.33
3	18	12.50	66	45.83
4	25	17.36	91	63.19
5	17	11.81	108	75.00
6	10	6.94	118	81.94
7	15	10.42	133	92.36
8	11	7.64	144	100.00

The next section of the results shows the counts graphically as a *bar chart*, which appears in Figure 3.1.

Figure 3.1: Bar Chart for Horse Racing Data

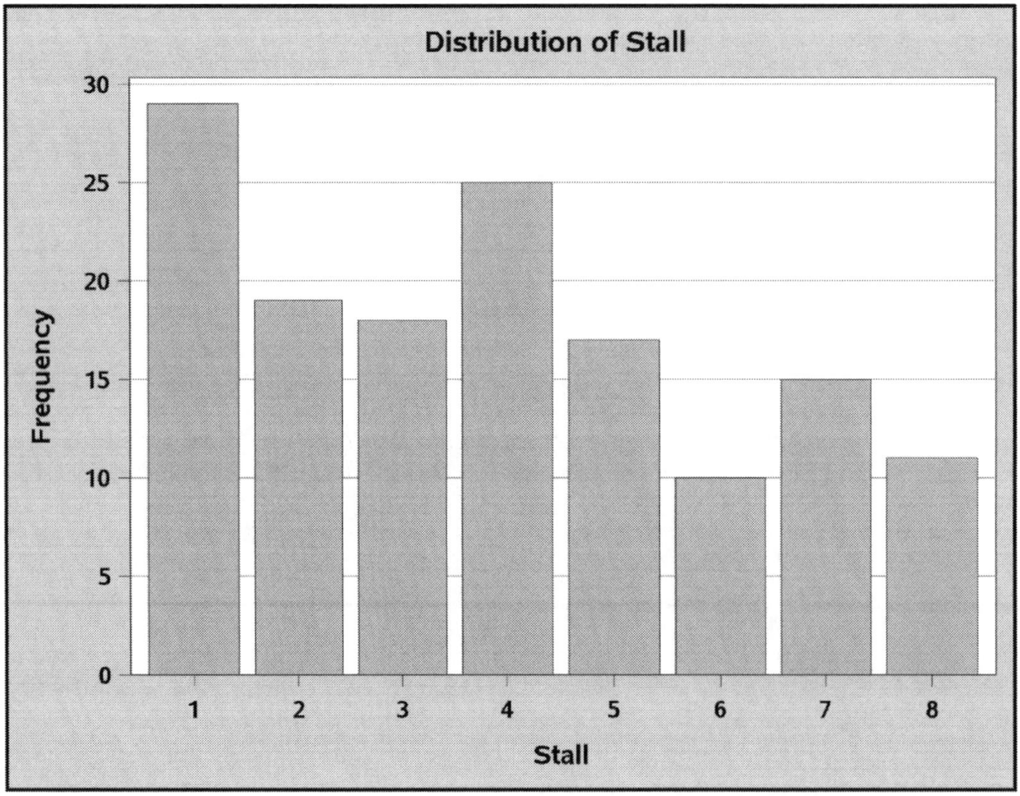

The same bar chart might also be produced separately by selecting **Tasks ▶ Graph ▶ Bar Chart** or, alternatively, as a *pie chart* as follows:

1. Open **Tasks ▶ Graph ▶ Pie Chart**.
2. Under **Data ▶ Data**, add **sasue.racestalls**.
3. Under **Data ▶ Roles ▶ Category variable**, add **stall**.
4. Click **Run**.

The resulting diagram is shown in Figure 3.2. It should be pointed out that despite their widespread popularity, both the general and scientific use of pie charts have been severely criticized (see Tufte, 1983, and Cleveland, 1994, for reasons). Both diagrams simply mirror what we previously gleaned from the percentages in Output 3.1, namely that there does appear to be a difference in the number of winners from each stall.

Figure 3.2: Pie Chart for Horse Racing Data

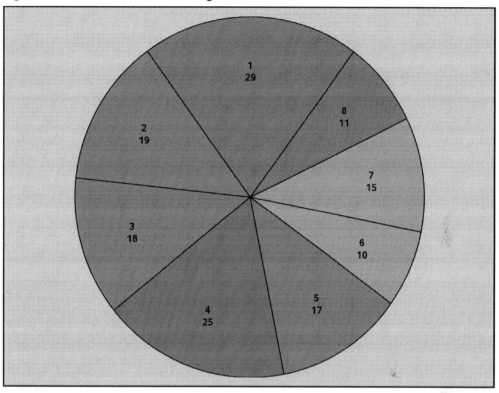

The bar chart often becomes more useful if the bars are arranged in ascending or descending order of frequency. This is not one of the standard options available within the Bar Chart task using the **Data** or **Options** tabs, but it is an option within the SGPLOT procedure that underlies the Bar Chart task. To utilise the option, we need to edit the code produced by the Bar Chart task:

1. Open **Tasks ▶ Graph ▶ Bar Chart**.
2. **Under Data ▶ Data**, add **sasue.racestalls**.
3. Under **Data ▶ Roles ▶ Category Variable**, add **stall**.
4. In the code pane, click the **Edit** button. The code pane then enlarges and you see the following code:

```
proc sgplot data=SASUE.RACESTALLS noautolegend;
    /*--Bar chart settings--*/
    vbar stall / name='Bar';

    /*--Response Axis--*/
    yaxis grid;
run;
```

5. Edit the VBAR statement to the following, taking care that the semicolon is at the end:

```
vbar stall / name='Bar' categoryorder=respasc;
```

6. Run the program.

The edited code could be saved as a code snippet by clicking on the **Snippet** button (📄) and typing a name, such as ordered bar chart.

The resulting plot is shown in Figure 3.3. We can now see clearly that starting stalls 1 to 4 produce many more winners than stalls 5 to 8, and starting stall 1 produces the highest number of winners of all eight starting stalls.

Figure 3.3: Ordered Bar Chart for Horse Racing Data

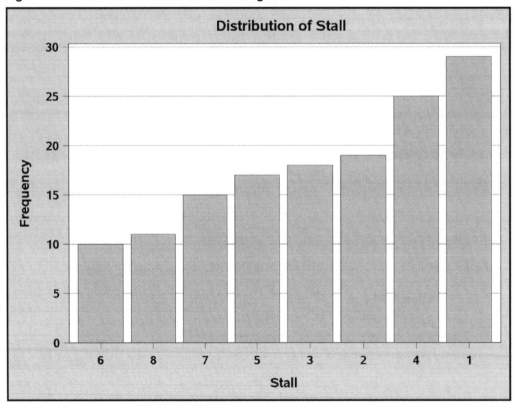

3.2.2 Chi-Square Goodness-of-Fit Test: Does Starting Stall Position Predict Horse Race Winners?

What would we expect the counts in Table 3.1 to look like if the starting stall does *not* affect the chances of a horse winning a race? Clearly, we would expect the number of winners from each stall to be approximately equal (random variation will stop them from being exactly equal). So here, our null hypothesis about the population of horse race winners is that there are an equal number of winners from each stall. In our sample of 144 winners, the counts do not appear to be consistent with the null hypothesis, but how can we assess the evidence against the null hypothesis formally? We begin by calculating the counts of winners in each stall that we might expect when we observe the results of 144 races, if the null hypothesis is true. We then compare these *expected values* with the observed values using what is known as the *chi-squared test statistic*. The expected values for each stall under the null hypothesis are simply 144/8=18, and the chi-squared statistic is then calculated as the sum of the square of each difference between the observed and expected value divided by the expected value. So, in detail, the required chi-squared test statistic is calculated thus:

$$\frac{(29-18)^2}{18} + \frac{(19-18)^2}{18} + \frac{(18-18)^2}{18} + \frac{(25-18)^2}{18} + \frac{(17-18)^2}{18} + \frac{(10-18)^2}{18} + \frac{(15-18)^2}{18} + \frac{(11-18)^2}{18}$$

If the null hypothesis is true, the chi-squared test statistic has a *chi-squared distribution* with 7 *degrees of freedom*. (Full details of the chi-square test are given in Altman, 1991.) To apply the test:

1. Return to the **One Way Frequencies task** (reopen it or repeat the steps to produce Output 3.1).
2. Under **Options ▶ Statistics ▶ Chi-square goodness-of-fit**, select **Asymptotic test**.
3. Click **Run**.

The results now include the section shown in Output 3.2 and Figure 3.4. The chi-squared statistic takes the value 16.3 with an associated *p*-value of 0.02. Consequently, there is evidence that the starting stall is a factor in determining the winning horse, as previously suggested by examination of the frequencies and the corresponding bar charts.

Ouput 3.2: Chi-Square Test for Horse Racing Data

Chi-Square Test for Equal Proportions	
Chi-Square	16.3333
DF	7
Pr > ChiSq	0.0222

Figure 3.4 demonstrates that the number of winners from stalls 1 to 4 are more than would be expected if the stall did not affect the race results, with stalls 5 to 8 leading to fewer winners than would be expected.

Figure 3.4 Deviation Plot for Horse Racing Stall

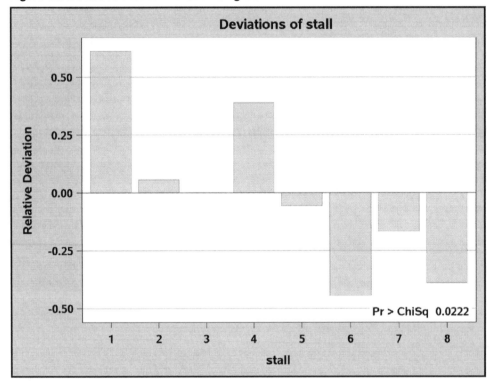

3.3 Two-By-Two Tables

Categorical variables having just two categories are often called *binary variables*, and when two such variables are cross-classified, we get the 2 x 2 table (or the 2 x 2 *contingency table*), ubiquitous particularly in medical studies but also in many other disciplines.

3.3.1 Chi-Square Test: Breast Self-Examination

Senie et al. (1981) report the results on asking women about how frequently they carried out breast self-examination. The data collected can be presented as a 2 x 2 table with one variable being the age of a woman, categorized as younger than 45 years or 45 years or older, and the other variable being how often a woman performed a self-examination of her breasts, categorized as monthly and occasionally/never. The cross-classification of the observations on the two variables are shown in Table 3.2.

Table 3.2: Data on Breast Self-Examination

Age	Monthly	Occasionally/Never	Total
Younger than 45	91	141	232
45 or older	259	705	964
Total	350	846	1196

We can use these data to determine whether there is any evidence that the population proportion of younger women (<45 years) who self-examine their breasts monthly differs from the corresponding proportion for older women (>=45 years) or, in other words, whether age and frequency of self-examination are related (that is, not independent of one another). The appropriate chi-square test for assessing a 2 x 2 table for independence is given explicitly in Der and Everitt (2013) and can be applied as follows:

1. Open **Tasks ▶ Statistics ▶ Table Analysis**.
2. Under **Data ▶ Data**, add **sasue.self_exam**.
3. Under **Data ▶ Roles**, add **agegroup** to the **Row variables** box and **frequency** to the **Column variables** box. Expand the **additional roles** and assign **num** as the **Frequency count**.
4. Under **Options ▶ Statistics**, note that **Chi-square statistics** is selected by default.
5. Under **Options ▶ Plots**, select **Suppress plots** to simplify the output.
6. Click **Run**.

The results are shown in Output 3.3. The chi-square statistic takes the value 13.8 with an associated *p*-value, which is very small. Clearly, age and frequency of self-examination of breasts are not independent of each other. We shall investigate the association in more detail in the next subsection. The other test statistics given in Output 3.3 are explained in detail in Everitt (1992). Fisher's exact test is dealt with below in Section 3.4.

Output 3.3: Testing for the Independence of Age and Frequency of Breast Examination

Table of agegroup by frequency			
agegroup	frequency		
Frequency	Monthly	Rarely	Total
Over45	259	705	964
Under45	91	141	232
Total	350	846	1196

Statistics for Table of agegroup by frequency

Statistic	DF	Value	Prob
Chi-Square	1	13.7936	0.0002
Likelihood Ratio Chi-Square	1	13.2376	0.0003
Continuity Adj. Chi-Square	1	13.2031	0.0003
Mantel-Haenszel Chi-Square	1	13.7821	0.0002
Phi Coefficient		-0.1074	
Contingency Coefficient		0.1068	
Cramer's V		-0.1074	

Fisher's Exact Test	
Cell (1,1) Frequency (F)	259
Left-sided Pr <= F	0.0002
Right-sided Pr >= F	0.9999
Table Probability (P)	<.0001
Two-sided Pr <= P	0.0003

Sample Size = 1196

3.3.2 Odds and Odds Ratios

When 2 x 2 tables are reported in the literature, the chi-square test described in the previous subsection is often supplemented by giving the value of what is known as the *odds ratio* of the table; in essence, this quantifies the degree of association between the two binary variables that form the table. Odds ratios are described in detail in Agresti (1996) and Der and Everitt (2013); here we content ourselves with a relatively brief account.

Returning to the breast self-examination data, we can estimate the probability that a woman younger than 45 rarely self-examines her breasts as 141/232=0.61 and the corresponding probability that she self-examines monthly is estimated as 91/232=0.39, which is, of course, simply 1 - 0.61. The ratio of these two estimated probabilities gives the estimated odds of rarely using self-examination to monthly examination (that is, the value 1.54); here the value is more than 1, indicating that a larger proportion of women younger than 45 tend to only rarely do self-examination than monthly self-examination.

Now for women aged 45 or older, the estimated odds calculated in the same way is 2.72, again indicating that for these older women, a smaller proportion do regular as opposed to occasional self-examination.

The ratio of the two estimated odds is our estimate of the odds ratio, namely 0.60. A population odds ratio of 1 corresponds to the independence of the two binary variables, so the chi-square test described above is also essentially a test that the population odds ratio has the value 1. But we can find a confidence interval for the odds ratio that enables us to say a little more about the strength of the relationship (if any) between the two variables (that is, to quantify the relationship, which is usually far more informative than simply quoting the value of the chi-square statistic). Details of how this confidence interval is calculated are given in Der and Everitt (2013). The odds ratio and its confidence interval can be obtained as follows:

1. Reopen the **Table Analysis** task. If it has been closed, repeat the previous example.
2. Under **Options** ▶ **Statistics**, select **odds ratio and relative risk**.
3. Click **Run**.

The relevant section of the output is shown in Output 3.4.

Output 3.4: Qdds Ratio for Age and Breast Examination Data

Odds Ratio and Relative Risks		
Statistic	**Value**	**95% Confidence Limits**
Odds Ratio	0.5692	0.4219	0.7680
Relative Risk (Column 1)	0.6850	0.5658	0.8292
Relative Risk (Column 2)	1.2033	1.0777	1.3435

Sample Size = 1196

From Output 3.4, we find the 95% confidence interval for the odds ratio is [0.42, 0.77]. This interval does not contain the value 1, so there is evidence that age and frequency of breast examination are not independent, which is, of course, in agreement with the result of the chi-square test. But from the confidence interval for the odds ratio, we can claim that the odds of rare rather than regular breast examination for younger women is between about 0.4 and 0.8 times the corresponding odds for older women. In other words, the odds are between 20% and 60% lower for women less than 45 years old. It appears that the message that self-examination of breasts is important has been taken on board by a considerably greater proportion of women under 45 rather than women over 45. (The relative risk terms in Output 3.4 are explained in Der and Everitt, 2013, and in Agresti, 1996.)

3.3.3 The Cochran-Mantel-Haenszel Test for Multiple Related 2 X 2 Tables

Agresti (1996) describes a set of data based on those in Liu (1992), which involves smoking and lung cancer in China. Part of these data are shown in Table 3.3.

Table 3.3: Chinese Smoking and Lung Cancer Data

City	Smoking	Lung Cancer		Odds Ratio
		Yes	No	
Beijing	Smokers	126	100	2.20
	Nonsmokers	35	61	
Shanghai	Smokers	908	688	2.14
	Nonsmokers	497	807	
Nanjing	Smokers	235	172	2.85
	Nonsmokers	58	121	
Zhengzhou	Smokers	182	156	1.58
	Nonsmokers	72	98	

As the main aim of this study is to investigate the relationship between smoking and lung cancer, it might be asked why not simply collapse the data over cities into a single 2 x 2 table and use the methods described previously to assess this relationship? The dangers of this procedure are well-known and collapsing in this way can, in some circumstances, generate spurious associations and, in others, mask a true relationship (see Everitt, 1992, for examples). Instead, we use the *Cochran-Mantel-Haenszel* (CMH) method; essentially, this approach tests that smoking behaviour and lung cancer are *conditionally* independent given the city. The method is described in detail in Der and Everitt (2013) and can be applied as follows:

1. Open **Tasks ▶ Statistics ▶ Table Analysis**.
2. Under **Data ▶ Data**, add **sasue.ca_lung**.
3. Under **Data ▶ Roles**, add **smoker** to the **Row variables**, add **cancer** to the **Column variables**, and **city** to the **Strata variables**. Expand the **additional roles** and assign **count** as the **Frequency count**.
4. Under **Options ▶ Statistics**, select **Cochran-Mantel-Haenszel statistics**. We could also select **odds ratio and relative risk** to calculate the separate odds ratios for each city, but as those are already available, we can omit it.
5. Under **Options ▶ Plots**, select **Suppress plots** to simplify the output.
6. Click **Run**.

The results are shown in Output 3.5.

Output 3.5: Results for the Cochran-Mantel-Haenszel Test Applied to the Chinese Smoking and Lung Cancer Data

Summary Statistics for smoker by cancer
Controlling for city

Cochran-Mantel-Haenszel Statistics (Based on Table Scores)				
Statistic	Alternative Hypothesis	DF	Value	Prob
1	Nonzero Correlation	1	144.5873	<.0001
2	Row Mean Scores Differ	1	144.5873	<.0001
3	General Association	1	144.5873	<.0001

Common Odds Ratio and Relative Risks				
Statistic	Method	Value	95% Confidence Limits	
Odds Ratio	Mantel-Haenszel	2.1450	1.8924	2.4313
	Logit	2.1437	1.8909	2.4302
Relative Risk (Column 1)	Mantel-Haenszel	1.4331	1.3525	1.5185
	Logit	1.4357	1.3551	1.5211
Relative Risk (Column 2)	Mantel-Haenszel	0.6677	0.6228	0.7158
	Logit	0.6707	0.6257	0.7189

Breslow-Day Test for Homogeneity of the Odds Ratios	
Chi-Square	4.8053
DF	3
Pr > ChiSq	0.1866

Total Sample Size = 4316

The result we are looking for is in the last part of the output—namely, the chi-square value of 4.8053, which has three degrees of freedom (one less than the number of tables) and an associated *p*-value of 0.1866. There is no evidence that the odds ratios for the relationship between smoking behaviour and cancer in the four cities considered differ from each other. Given this result, we can use the data from all cities to estimate the odds ratio common to all cities; details of the calculation are given in Der and Everitt (2013) and the value we want is to be found in part of Output 3.5: 2.15 with 95% confidence interval [1.89, 2.43]. The odds of cancer versus no cancer for smokers equals 2.15 times the corresponding odds for non-smokers; the confidence interval shows that the estimated odds are between 89% and 143% higher for smokers.

3.4 Larger Cross-Tabulations

In an investigation of brain tumours, the type and site of the tumour in 141 individuals were noted. The three possible types were: A, benign tumours; B, malignant tumours; and C, other cerebral tumours. The sites concerned were: I, frontal lobes; II, temporal lobes; and III, other cerebral areas. The data are shown in Table 3.4. Do these data give any evidence that some types of tumours occur more frequently at particular sites (that is, that there is an association between the categorical **type** and **site** variables)?

Table 3.4: Data on Type and Site of Brain Tumours

1 III A	44 II A	87 I A	130 III B
2 III C	45 III B	88 II A	131 III B
3 II A	46 II A	89 I A	132 III A
4 I A	47 II A	90 III A	133 III C
5 III A	48 III A	91 III A	134 III C
6 III C	49 I B	92 III B	135 III B
7 I A	50 III C	93 III C	136 III A
8 I A	51 III B	94 I A	137 I A
9 III A	52 III C	95 III A	138 I B
10 III A	53 III A	96 II A	139 III B
11 III A	54 I A	97 I B	140 II A
12 I A	55 III C	98 II B	141 I A
13 III A	56 III C	99 II A	
14 III B	57 III A	100 III B	
15 III A	58 III A	101 III B	
16 III B	59 III B	102 III C	
17 II A	60 III A	103 I A	
18 III A	61 II A	104 III C	
19 I B	62 III A	105 III A	
20 III C	63 III A	106 III A	
21 I A	64 I A	107 II A	
22 III A	65 II C	108 I C	
23 III A	66 III B	109 III A	
24 III A	67 III A	110 III C	
25 III A	68 I A	111 II A	
26 III B	69 I A	112 III B	
27 III B	70 II A	113 III C	
28 II A	71 III B	114 II A	
29 I B	72 I C	115 I B	
30 III B	73 II A	116 I B	
31 II C	74 III C	117 II B	
32 III A	75 I A	118 III B	
33 II A	76 II A	119 II A	
34 II A	77 III A	120 III C	
35 I A	78 III C	121 I C	
36 III B	79 III A	122 I A	
37 II B	80 I A	123 I C	
38 II B	81 II A	124 I A	
39 I B	82 I A	125 III A	
40 III B	83 III B	126 III A	
41 I C	84 II C	127 III B	
42 I A	85 I C	128 III B	
43 I B	86 I A	129 III A	

3.4.1 Tabulating the Brain Tumour Data Into a Contingency Table

For the data about brain tumours in Table 3.9, we can cross-classify the observations to give what is know as a 3 x 3 *contingency table* showing the counts in all nine possible combinations of the type and site of tumour categories. The original data are in a SAS data set in the Sasue library, **tumours**:

1. Open **Tasks ▶ Statistics ▶ Table Analysis**.
2. Under **Data ▶ Data**, add **sasue.tumors**.
3. Under **Data ▶ Roles**, add **site** to the **Row variables** box and **type** to the **Column variables** box.
4. Under **Options ▶ Plots**, select **Suppress plots**.
5. Under **Options ▶ Statistics**, note that **Chi-square statistics** is selected by default.
6. Click **Run**.

The first part of the results is the contingency table shown in Output 3.6.

Output 3.6: Brain Tumour Data After Cross-Classification

Table of site by type				
site	type			
Frequency	A	B	C	Total
I	23	9	6	38
II	21	4	3	28
III	34	24	17	75
Total	78	37	26	141

3.4.2 Do Different Types Of Brain Tumours Occur More Frequently at Particular Sites? Chi-Squared Test

We are now interested in assessing the null hypothesis that the site of tumour and the type of tumour are *independent*. Independence implies that the probabilities of the tumour types are the same at all sites. More explicitly, independence implies that the probability of a patient having a tumour of a particular type at a particular site is simply the product of the probability of this type of tumour multiplied by the probability of a tumour at this site. We can estimate both the probability of the type of tumour and the probability of a tumour at a particular site by simply dividing the appropriate *marginal total* by the number of observations. So, for example, the estimate of the probability of being a type A tumour is 78/141=0.553, and the estimate of a tumour being at site I is 38/141=0.270. So, if the null hypothesis of independence is true, the estimate of the probability of a patient having an A type tumour at site I is 0.553 x 0.270=0.149. So, under the assumption of independence, the expected count in the type A, site I cell of the contingency table is 141 x

0.149=21.0. In the same way, we can calculate the expected values for all the other cells in the table, and these can then be compared with the observed values by means of the chi-square statistic. For a contingency table with r rows and c columns, the chi-squared test of independence has $(r-1)(c-1)$ degrees of freedom, where r is the number of rows of the table and c is the number of columns. In the tumour example, both r and c have the value 3, so the chi-squared statistic will have 4 degrees of freedom. (Full details of the chi-squared test of independence in contingency tables are given in Everitt, 1992.)

As noted above, for the Table Analysis task, **chi-square statistics** is available within **Options ▶ Statistics** and is selected by default.

The result is shown in Output 3.7. Here, the chi-square test statistic takes the value 7.8 and has an associated p-value of 0.098; there is no strong evidence against the hypothesis that type and site of tumour are independent. The result implies that the observed values in Output 3.6 do not differ greatly from the corresponding values to be expected if tumour site and type are independent. (We will not describe the other terms in Output 3.7; interested readers are referred to Everitt, 1992.)

Output 3.7: Chi-Squared Test of Independence for Brain Tumour Data

Statistic	DF	Value	Prob
Chi-Square	4	7.8441	0.0975
Likelihood Ratio Chi-Square	4	8.0958	0.0881
Mantel-Haenszel Chi-Square	1	2.9753	0.0845
Phi Coefficient		0.2359	
Contingency Coefficient		0.2296	
Cramer's V		0.1668	

The chi-square test statistic and associated p-value simply summarise the evidence against the null hypothesis of independence. But we can often go further by making cell-by-cell comparisons of the differences of observed and estimated expected frequencies under independence, differences usually labelled by statisticians as *residuals*. Unfortunately, these very simple residuals are not entirely satisfactory because large values tend to occur for cells that have larger expected values; consequently, they might be misleading. A more appropriate residual would be

$$(\text{Observed-Estimated Expected})/\sqrt{\text{Estimated Expected}}$$

these are usually know as *Pearson residuals* but for the reasons given in Everitt (1992), the use of Pearson residuals for detailed examination of a contingency table can often give conservative indications of cells that do not fit the independence hypothesis. More suitable residuals are given by Haberman (1973); these are calculated as

$$\text{Pearson residual}/\sqrt{(1-\frac{\text{row total}}{\text{sample size}})(1-\frac{\text{column total}}{\text{sample size}})}$$

These are known as *adjusted residuals* or *standardized residuals.*

The first of these three types of residuals is available under **Options ▶ frequency table ▶ Frequencies** by selecting **Deviation.** To produce the other two types of residual requires us to edit the code. We will illustrate all three as follows;

1. Return to the Tables Analysis task above.
2. Under **Options ▶ Frequency Table ▶ Frequencies**, select **Observed, Expected,** and **Deviation.**
3. In the code pane, click the **Edit** button. The code window should include the following:

    ```
    proc freq data=SASUE.TUMORS;
        tables (site)*(type) / chisq expected deviation nopercent
    norow nocol nocum plots=none;
    ```

    After `plots=none` but before the semicolon, type **crosslist(pearsonres stdres)**.

4. Run the program.

The resulting cross-tabulation now appears as shown in Output 3.8.

The table is now in the list format, with one row per cell of the original table, showing the observed and estimated expected counts and the three types of residuals.

Output 3.8: Brain Tumour Data Showing Deviation, Pearson, and Standardized Residuals Per Cell

				Table of site by type		
site	type	Frequency	Expected	Deviation	Std Residual	Pearson Residual
I	A	23	21.0213	1.9787	0.7554	0.4316
	B	9	9.9716	-0.9716	-0.4192	-0.3077
	C	6	7.0071	-1.0071	-0.4929	-0.3805
	Total	38				
II	A	21	15.4894	5.5106	2.3399	1.4002
	B	4	7.3475	-3.3475	-1.6063	-1.2350
	C	3	5.1631	-2.1631	-1.1775	-0.9520
	Total	28				
III	A	34	41.4894	-7.4894	-2.5425	-1.1627
	B	24	19.6809	4.3191	1.6569	0.9736
	C	17	13.8298	3.1702	1.3797	0.8525
	Total	75				
Total	A	78				
	B	37				
	C	26				
	Total	141				

Looking at the standardized residuals, we see that the greatest departures from independence occur for type A tumours in the second site, with more tumours than would be expected under independence, and for the same type of tumour again in the third site, where there are fewer tumours than expected.

3.5 Fisher's Exact Test

Mann (1981) reports a study carried out to investigate the causes of jeering or baiting behaviour by a crowd when a person is threatening to commit suicide by jumping from a high building. A hypothesis is that baiting is more likely to occur in warm weather. Mann classified 21 accounts of threatened suicide by two factors, the time of the year and whether baiting occurred. The (classified) data are given in Table 3.5 and the question is whether they give any evidence to support the warm weather hypothesis. (The data come from the northern hemisphere, so the months June to September are the warm months.)

Table 3.5: Crowd Behaviour at Threatened Suicides

	Baiting	**Nonbaiting**
June-September	8	4
October-May	2	7

3.5.1 How Is Baiting Behaviour at Suicides Affected by Season? Fisher's Exact Test

The chi-squared test carried out in the previous section for the brain tumour data above depends on knowing that the test statistic has a chi-squared distribution if the null hypothesis of independence is true; this allows *p*-values to be found. But what was not mentioned previously is that the chi-squared distribution is only appropriate under the assumption that the expected values are not too small. Such a term is almost as vague as asking how long is a piece of string and has been interpreted in a number of ways. Most commonly, it has been taken as meaning that the chi-squared distribution is only appropriate if all the expected values are five or more. Such a rule is widely quoted but appears to have little mathematical or empirical justification over, say, a one-or-more rule.

Nevertheless, for contingency tables based on small sample sizes, the usual form of the chi-squared test for independence might not be strictly valid, although it is often difficult to predict *a priori* whether a given data set might cause problems. But there might be occasions where it is advisable to consider another approach that is available and that is a test that does not depend on the chi-squared distribution at all. Such *exact* tests of independence for a general *r* x *c* contingency table are computationally intensive and, until relatively recently, the computational difficulties have severely limited their application. But within the last 10 years, the advent of fast algorithms and the availability of inexpensive computing power have considerably extended the bounds where exact tests are feasible. Details of the algorithms for applying exact tests are outside the level of this text and interested readers are referred to Mehta and Patel (1986) for a full exposition. But for a table in which both *r* and *c* equal two, there is an exact test that has been in use for decades--namely *Fisher's exact test,* a test that is described in Everitt (1992). Fisher's test is produced by default as part of chi-square tests for a 2 x 2 contingency table. (For larger tables, it is available as an option.)

The data on baiting behaviour at suicides provides us with an example of how to apply Fisher's exact test for a 2 x 2 table and also serves to illustrate how to analyze data that is in the form of a table rather than individual observations. We begin by opening the data set, **baiting**, to examine its contents:

1. Select **Libraries ▶ sasue ▶ baiting**.
2. Drag across or double-click.

The view of the opened data set is shown in Display 3.1. It has one row per cell and a column each for the number in the cell; whether there was baiting, and whether the season was warm or cool.

Display 3.1: Baiting Data in Tabular Form

Total rows: 4   Total columns: 3

	baiting	season	count
1	B	warm	8
2	B	cool	4
3	N	warm	2
4	N	cool	7

To apply the chi-square and Fisher's exact test:

1. Open **Tasks ▶ Statistics ▶ Table Analysis**.
2. Enter **sasue.baiting** in the data box, add **Season** to the **Row Variables** box and baiting to the Column Variables box.
3. Expand Additional Roles and add **count** to the **Frequency count** box.
4. Under the **Options ▶ Statistics**, we see that **chi-square statistics** is selected by default, but under **Exact Test**, **Fisher's exact test** is not selected. We could select it, but for 2 x 2 tables, it is applied by default.

5. Under **Options ▶ Plots,** select **Suppress plots.**
6. Click **Run.**

The result is shown in Output 3.9. The cross-tabulation is not reproduced exactly as entered – the categories of **season** and **baiting** are in alphabetical order.

The *p*-value from Fisher's test is 0.0805. There is no strong evidence of crowd behaviour being associated with time of year of threatened suicide, but it has to be remembered that the sample size is low and the test lacks power. (Carrying out the usual chi-squared test on these data gives a *p*-value of 0.0436, a considerable difference from the value for the exact test, suggesting there *is* evidence of an association between crowd behaviour and time of year of threatened suicide.)

Output 3.9: Analysis of Baiting and Suicide Data

Table of season by baiting			
season	baiting		
Frequency	B	N	Total
cool	4	7	11
warm	8	2	10
Total	12	9	21

Statistics for Table of season by baiting

Statistic	DF	Value	Prob
Chi-Square	1	4.0727	0.0436
Likelihood Ratio Chi-Square	1	4.2535	0.0392
Continuity Adj. Chi-Square	1	2.4858	0.1149
Mantel-Haenszel Chi-Square	1	3.8788	0.0489
Phi Coefficient		-0.4404	
Contingency Coefficient		0.4030	
Cramer's V		-0.4404	
WARNING: 50% of the cells have expected counts less than 5. Chi-Square may not be a valid test.			

Fisher's Exact Test	
Cell (1,1) Frequency (F)	4
Left-sided Pr <= F	0.0563
Right-sided Pr >= F	0.9942
Table Probability (P)	0.0505
Two-sided Pr <= P	0.0805

Sample Size = 21

3.5.2 Fisher's Exact Test for Larger Tables

Fisher's test can also be applied to tables with more than two rows and/or more than two columns. Exact tests for contingency tables in which the counts are small have been developed by Mehta and Patel (1986) and, to illustrate, we shall use the data shown in Table 3.6; these data give the distribution of oral cancers found in house-to-house surveys in three geographic regions of rural India by the site of the lesion and by the region. The counts in the table are very small, so we apply the exact test. The data are in tabular form in the data set **lesions**:

1. Open **Tasks ▶ Statistics ▶ Table Analysis**.
2. Under **Data ▶ Data**, add **sasue.lesions**.
3. Under **Data ▶ Roles**, add **site** to the **Row variables** and **region** to the **Column variables**.
4. Under **Data ▶ Roles ▶ Additional Roles**, add **n** as the **Frequency count**.
5. Under **Options ▶ Statistics**, note that **chi-square statistics** is selected by default, but now under **Exact Test**, select **Fisher's exact test**.
6. Under **Options ▶ Plots**, select **Suppress plots**.
7. Click **Run**.

Table 3.6: Data on Oral Lesions in Different Regions of India

Site of lesion	Region		
	Keral	Gujarat	Andhra
Buccal Mucosa	8	1	8
Labial Mucosa	0	1	0
Commissure	0	1	0
Gingiva	0	1	0
Hard palate	0	1	0
Soft palate	0	1	0
Tongue	0	1	0

Site of lesion	Region		
	Keral	**Gujarat**	**Andhra**
Floor of mouth	1	0	1
Alveolar ridge	1	0	1

The results are shown in Output 3.10. The *p*-value from Fisher's test is 0.01, which indicates a strong association between site of lesion and geographic region. For comparison, the chi-squared statistic for these data takes the value 22.01 with 14 degrees of freedom and a *p*-value of 0.14, suggesting that there is no evidence against the hypothesis of independence of site and region. Here the data are so sparse that the usual chi-square test is misleading.

Output 3.10: Statistics for Table of Site by Region

Statistics for Table of site by region

Statistic	DF	Value	Prob
Chi-Square	16	22.0992	0.1400
Likelihood Ratio Chi-Square	16	23.2967	0.1060
Mantel-Haenszel Chi-Square	1	0.0000	1.0000
Phi Coefficient		0.9047	
Contingency Coefficient		0.6709	
Cramer's V		0.6397	

WARNING: 93% of the cells have expected counts less than 5. Chi-Square may not be a valid test.

Fisher's Exact Test	
Table Probability (P)	<.0001
Pr <= P	0.0101

Sample Size = 27

3.6 McNemar's Test Example

The data in Table 3.7 (taken from Agresti, 1996) arise from a sample of juveniles convicted of felony in Florida in 1987. *Matched pairs* of offenders were formed using criteria such as age and number of previous offences. For each pair, one subject was handled in the juvenile court and the other was transferred to the adult court. Whether the juvenile was rearrested by the end of 1988 was then noted. Here the question of interest is whether the population proportions rearrested are identical for the adult and juvenile courts.

Table 3.7 Rearrests of Juvenile Felons by Type of Court in Which They Were Tried

	Juvenile Court	
Adult Court	**Rearrest**	**No rearrest**
Rearrest	158	515
No rearrest	290	1134

3.6.1 Juvenile Felons: Where Should They Be Tried? McNemar's Test

The chi-squared test on categorical data described previously assumes that the observations are independent of one another. But the data on juvenile felons in Table 3.15 arise from *matched pairs*, so they are *not* independent. The counts in the corresponding 2 x 2 table of the data refer to the pairs; so, for example, in 158 of the pairs of offenders, *both* members of the pair were rearrested. To test whether the rearrest rate differs between the adult and juvenile courts, we need to apply what is known as *McNemar's test*. The test is described in Everitt (1992). The juvenile offenders data are in the data set **sasue.arrests**, again in tabular form as shown in Display 3.2. (Open **Libraries ▶ Sasue ▶ Arrests** to see this.) To apply *McNemar's test*:

1. Open **Tasks ▶ Statistics ▶ Table Analysis**.
2. Under **Data ▶ Data**, add **sasue.arrests**.
3. Under **Data ▶ Roles**, add **adult** to the **Row variables** and **juvenile** to the **Column variables**.
4. Expand **Additional Roles** and add **num** to the **Frequency count** box.
5. Under **Options ▶ Statistics**, deselect **chi-square statistics** and select **Measures of agreement**.
6. Under **Options ▶ Plots**, select **Suppress plots**.
7. Click **Run**.

Display 3.2: Rearrest Data for Juvenile Felons

Total rows: 4   Total columns: 3

	adult	juvenile	num
1	y	y	158
2	y	n	515
3	n	y	290
4	n	n	1134

The result is shown in Output 3.11. The test statistics take the value 62.89, with an extremely small associated *p*-value. There is very strong evidence that the type of court and the probability of rearrest are related. It appears that trial in a juvenile court is less likely to result in rearrest.

Output 3.11: McNemar's Test for Juvenile Crime Data

Table of adult by juvenile

adult	juvenile		
Frequency	n	y	Total
n	1134	290	1424
y	515	158	673
Total	1649	448	2097

Statistics for Table of adult by juvenile

McNemar's Test	
Statistic (S)	62.8882
DF	1
Pr > S	<.0001

3.7 Exercises

Exercise 3.1: Crash Data

The **crash** data set lists fictitious counts of fatal air crashes in Australia by quarter over a 20-year period. Assess the hypothesis that the accident rates are uniform across these four quarters:

Jan	April	July	October
12	8	7	8

Exercise 3.2: Fear Data

One hundred American citizens were surveyed and asked to identify which of five items were most fearful to them. The results are given in the **fear** data set. Test whether sex and greatest fear are independent of each other.

	Public Speaking	Heights	Insects	Financial Problems	Sickness/Death
Male	12	5	4	17	10
Female	11	15	10	4	12

Exercise 3.3: Suicidal Data

In a broad sense, psychiatric patients can be classified as psychotics or neurotics. A psychiatrist whilst studying the symptoms of a random sample of 20 patients from each type found that whereas six patients in the neurotic group had suicidal feelings, only two in the psychotic group suffered in this way. Is there any evidence of an association between the type of patient and suicidal feelings?

	Psychotics	Neurotics
Suicidal feelings	2	6
No suicidal feelings	18	14

The data are in the **suicidal** data set.

Exercise 3.4: Cancer Data

The data in the **cancer** data set arise from an investigation of the frequency of exposure to oral conjugated estrogens among 183 cases of endometrial cancer. Each case was matched on age, race, date of admission, and hospital of admission to a suitable control not suffering from cancer. Is there any evidence that use of oral conjugated estrogens is associated with endometrial cancer?

		Controls	
		Used	**Not Used**
Cases	Used	12	43
	Not used	7	121

Exercise 3.5: Death Data

Van Belle et al. (2004) report on a study by Peterson et al. (1979) in which the age at death of children who died from various causes was recorded. The data were taken from death records in a region of the USA. The resulting cross-classification of age at death and cause of death is as follows and is in the dataset infdeath:

			Age of Death		
Cause of Death	**0 Days**	**1-6 Days**	**2-4 Weeks**	**5-26 Weeks**	**27-51 Weeks**
1	19	51	7	0	0
2	68	191	46	0	3
3	105	60	7	4	2
4	104	34	3	0	0
5	115	105	17	2	0
6	79	101	72	75	32
7	7	38	36	43	18
8	0	0	24	274	24
9	60	51	28	58	35

Investigate the relationship (if any) between the cause of death and the age at death.

Exercise 3.6: Victim Blaming Data

Howell (2002) describes a study reported in Pugh (1983) involving investigation of the "blaming the victim" phenomenon in prosecutions for rape. The data (in the 'rapes' dataset) extracted by Howell from Pugh's detailed report is as follows:

	Guilty Verdict		
	Yes	**No**	**Total**
Low	153	24	177
High	105	76	181
Total	258	100	358

The low and high categories are based on Pugh's assessment of how strongly the defence alleged that the victim was somewhat partially guilty for the alleged rape.

Do the data give any evidence that how the defence conducts their case in respect to blaming the victim affects the final verdict?

Chapter 4: Bivariate Data: Scatterplots, Correlation, and Regression

4.1 Introduction

When two observations or measurements are made on each member of a sample, we have what is generally termed *bivariate data*. In this chapter, we show how to construct informative graphical displays of such data and how to quantify the relationship between the two variables in such data sets. The statistical topics to be covered in this chapter are:

- Plotting the data: Scatter plots
- Assessing the strength of a linear relationship: Correlation coefficients
- Fitting a line to the data: Simple linear regression

4.2 Scatter Plots, the Correlation Coefficient, and Simple Linear Regression

The data in Table 4.1 show the heights (in cm) and resting pulse rates (beats per minute) for a sample of hospital patients. The main question of interest is whether there is any relationship between height and pulse rate.

Table 4.1: Height (cm) and Resting Pulse (bpm) Data

Height	Pulse	Height	Pulse	Height	Pulse
160	68	170	80	168	90
167	80	148	82	178	80
162	84	175	76	182	76
175	80	160	84	167	80
185	80	153	70	170	84
162	80	185	80	160	80
173	92	165	82	182	80
167	92	165	84	168	80
170	80	172	116	155	80
170	80	185	80	175	104
163	80	163	95	168	80
158	80	177	80	180	68
157	80	165	76	175	84
160	78	182	100	145	64
170	90	162	88	170	84
177	80	172	90	175	72
166	72	177	90		

4.2.1 The Scatter Plot

The height and resting pulse data set is termed *bivariate* because two variables are measured for each individual. The separate variables in the data set can, of course, each be summarized and graphed using the methods described in Chapter 2, but of more importance and interest for bivariate data is to describe and graph the data in a way that lends insights about how the two variables are related. Let's begin by looking at the most commonly used graphic for bivariate data—namely, the *scatter plot*, an *xy* plot of the two variables, which has been in use since at least the 18th century and has many virtues. Indeed, according to Tufte (1983):

The relational graphic—in its barest form the scatter plot and its variants—is the greatest of all graphical designs. It links at least two variables encouraging and even imploring the viewer to assess the possible causal relationship between the plotted variables. It confronts causal theories that x causes y with empirical evidence as to the actual relationship between x and y.

The data giving heights and resting pulse rates shown in Table 4.1 is in the data set, **resting**. To produce a scatter plot:

1. Open **Tasks ▶ Graph ▶ Scatter Plot**.
2. Under **Data ▶ Data**, add **sasue.resting**.
3. Under **Data ▶ Roles, add height** as the **X variable** and **pulse** as the **Y variable**.
4. Under **Options ▶ X axis**, deselect **Show grid line**. Do the same for **Y axis**.
5. Click **Run**.

The resulting plot is shown in Figure 4.1 and suggests that increasing height is generally (although not universally) associated with an increase in resting pulse and that the relationship between the two variables is, approximately at least, *linear*, and it might be able to be described satisfactorily by a straight line (see later).

Figure 4.1: Scatter Plot of Pulse Against Height

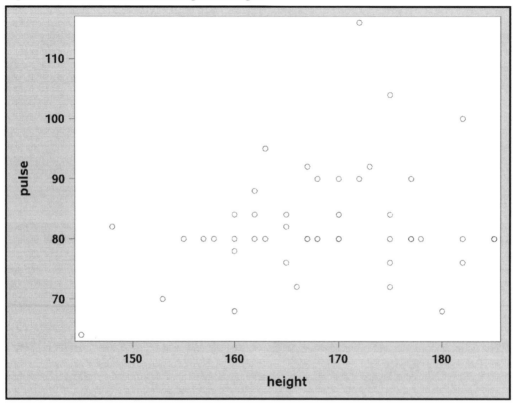

4.2.2 The Correlation Coefficient

How can we summarize and quantify any relationship between two variables indicated in the scatter plot of the two variables in a single number? What is needed is to measure the *correlation* between the two variables using a *correlation coefficient*. For two continuous variables, we can use *Pearson's correlation coefficient*, also known as the *product-moment correlation coefficient*.

The product-moment correlation coefficient is the ratio of the sum of products of differences of each variable from its mean divided by the square roots of the two sums of squares about the mean. (For more details about how the coefficient is calculated, see Altman, 1991.) The product-moment coefficient takes values between -1 and 1. Negative values indicate that large values of *x* are associated with small values of *y* and vice-versa. Positive values indicate the reverse. The correlation coefficient has a maximum value of +1 when the points in the scatter plot all lie exactly on a straight line and the variables are positively correlated. The correlation coefficient has a minimum of -1 when all the points lie exactly on a straight line and the variables are negatively correlated. When the correlation coefficient is 0, the variables are said to be *uncorrelated*.

In essence, the correlation coefficient is a measure of how closely the points in the scatter plot are to a straight line--it measures the *linear relationship* between two variables; non-linear relationships might be missed or underestimated by it. For example, Figure 4.2 shows a perfect non-linear relationship between two variables for which the correlation coefficient takes the value 0, and Figure 4.3 shows another perfect non-linear relationship for which the coefficient is not 1. The examples in Figures 4.2 and 4.3 demonstrate the need to use the scatter plot alongside the correlation coefficient when assessing relationships between variables. Use of the correlation coefficient alone is not sufficient and can lead to misinterpretation of the data.

Figure 4.2: Perfect Non-Linear Relationship Between Two Variables for Which the Correlation Coefficient Is Almost 0

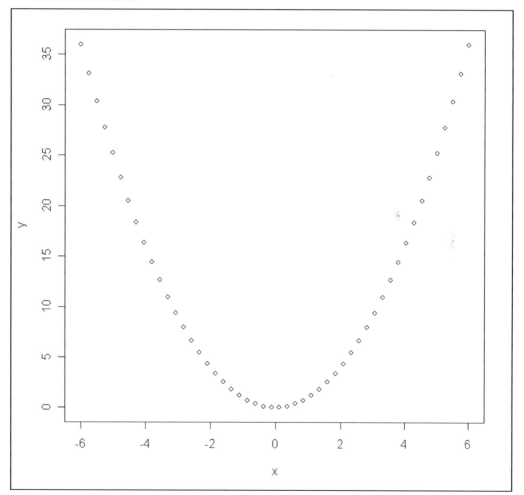

Figure 4.3: Perfect Non-Linear Relationship Between Two Variables for Which the Correlation Coefficient Is Nevertheless Not 1

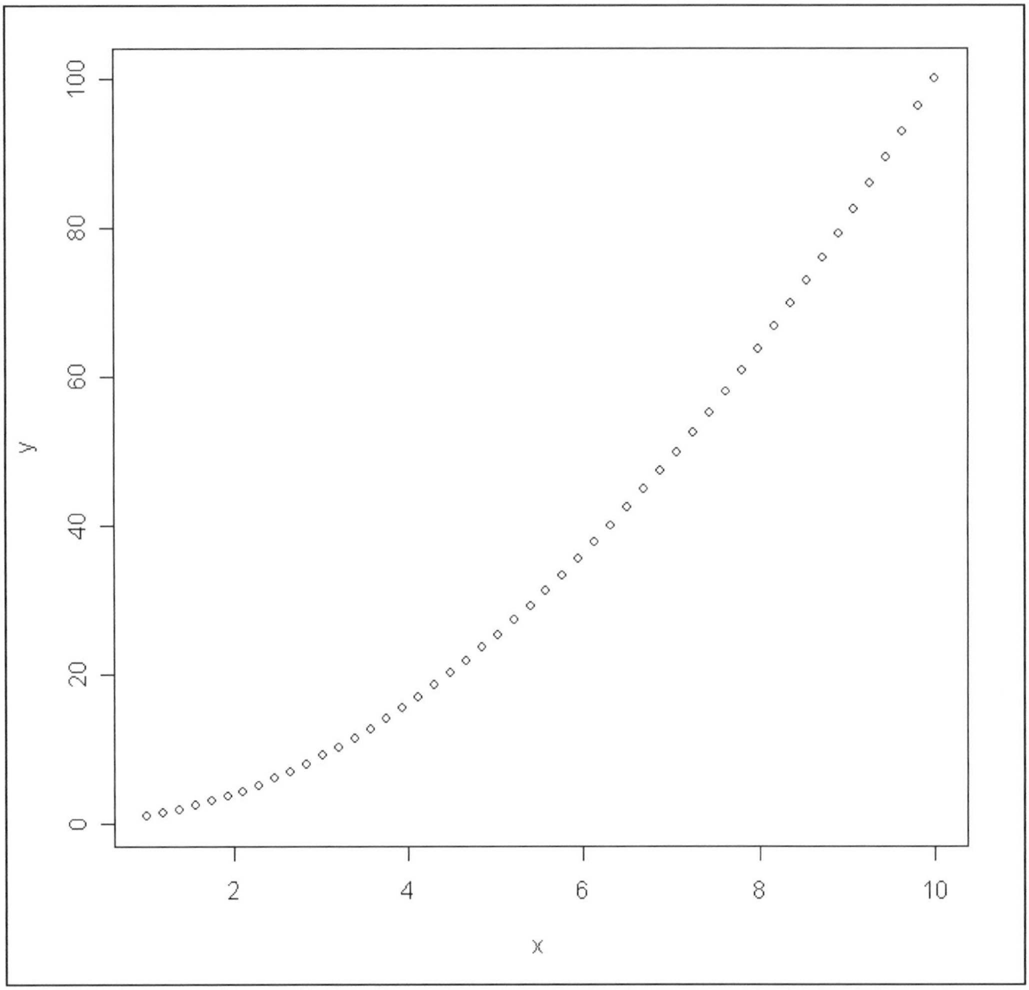

To calculate the product-moment correlation coefficients for height and resting pulse rate data:

1. Open **Tasks ▶ Statistics ▶ Correlation Analysis**.
2. Under **Data ▶ Data**, add **sasue.resting**.
3. Under **Data ▶ Roles**, add both variables, **height** and **pulse**, to the **Analysis variables** box.
4. Click **Run**.

The results are shown in Output 4.1.

Output 4.1: Correlation Coefficients for the Height and Resting Pulse Rate Data

Pearson Correlation Coefficients, N = 50 Prob > |r| under H0: Rho=0		
	height	pulse
height	1.00000	0.21822 0.1279
pulse	0.21822 0.1279	1.00000

The correlation between height and resting pulse is 0.22, which indicates a relatively weak positive association between the two variables. A correlation coefficient calculated from a sample of observations is an *estimate* of the corresponding value in the population (in the same way that the sample mean is an estimate of the population mean--see Chapter 2). Consequently, we may want to use the sample correlation as the basis of a test of some hypothesis about the population correlation. The most common hypothesis of interest is that the population value is 0 (that is, there is no linear relationship between the two variables). Under the hypothesis of no linear relationship,

a suitable test statistic is $t = r\sqrt{\dfrac{n-2}{1-r^2}}$ where n is the sample size and r is the sample correlation

coefficient. If the hypothesis of 0 population correlation is true, the statistic is known to have a Student's t-distribution with n-2 degrees of freedom. The result of the test is labelled **Prob > \|r\| under H0: Rho=0** in the results in Output 4.1. So for height and resting pulse with a p-value of 0.13, there is no evidence that the two variables are related; the population correlation between the two variables may well be 0. (This test for 0 population correlations assumes that the two variables have what is known as a *bivariate normal distribution*--see Everitt and Skrondal, 2010, but this assumption is rarely investigated.)

4.2.3 Simple Linear Regression

Rather than simply measuring the correlation between two variables, we would often like to derive an equation that links one variable to the other and might, in some situations, be used for *predicting* the values of one variable from the values of the other. And if such an equation can be derived, it is often also useful to add it to the scatter plot of the two variables to highlight their relationship. Most commonly, we wish to find the straight line that best fits the observed data. Fitting a straight line involves *simple linear regression* and *least squares estimation*, both of which are described in detail in Altman (1991). But, essentially, we postulate the following model for the data and then estimates the model's two *parameters*: α (the intercept of the line) and β (the slope of the line):

$$y_i = \alpha + \beta x_i + \varepsilon_i$$

In the model above, x_i, y_i represents the observed values of the two variables for the ith subject in the sample of observations, and ε_i represents the error (that is, the amount by which y_i differs from its value as predicted by the model--namely, $\alpha + \beta x_i$). The formulae for the sample estimates of α and β are given explicitly in Altman (1991). In this model, the error terms are assumed to have a normal distribution with constant variance (that is, the variance of the y-variable does not depend on the x-variable values).

We can fit the simple linear regression model to the height and resting pulse rate data and obtain an informative plot, as follows:

1. Open **Tasks** ▶ **Statistics** ▶ **Linear Regression**.
2. Under **Data** ▶ **Data**, enter **sasue.resting**.
3. Under **Data** ▶ **Roles**, add **pulse** as the **Dependent variable** and **height** to the **Continuous variables box**.
4. Under **Model** ▶ **Model Effects**, select **height** and click the **Add** button.
5. Under **Options** ▶ **Plots** ▶ **Scatter Plots**, select **Fit plot for a single continuous variable** and deselect the other plots.
6. Click **Run**.

The tabular results are shown in Output 4.2 and the plot in Figure 4.4.

Output 4.2: Results of Fitting a Simple Linear Regression Model to the Height and Pulse Rate Data

Number of Observations Read	50
Number of Observations Used	50

Analysis of Variance					
Source	DF	Sum of Squares	Mean Square	F Value	Pr > F
Model	1	186.32129	186.32129	2.40	0.1279
Error	48	3726.17871	77.62872		
Corrected Total	49	3912.50000			

Root MSE	8.81072	R-Square	0.0476
Dependent Mean	82.30000	Adj R-Sq	0.0278
Coeff Var	10.70561		

Parameter Estimates					
Variable	DF	Parameter Estimate	Standard Error	t Value	Pr > |t|
Intercept	1	46.90693	22.87933	2.05	0.0458
height	1	0.20977	0.13540	1.55	0.1279

The first part of Output 4.2 gives an *analysis of variance table* (see Chapter 5), in which the variation in the *y* variable is partitioned into a part due to the fitted model and a part due to the error term in the model. The associated F-test (again, see Chapter 5) gives a test of the hypothesis that the population value of the slope is 0 ($H_0 : \beta = 0$). Here the *p*-value associated with the F-test is 0.13, so there is no evidence of a non-0 slope (note that the *p*-value is the same as the previously described test for 0 correlation between the two variables; the two tests are, of course, equivalent).

The most important term in the second part of Table 4.2 is *R-square*, which is the square of the correlation between the observed values of the response variable and the values of the response variable predicted by the fitted model. R-square gives the variance in the response variable *y* that is explained by the *x* variable. Here, the R-square value of 0.0476 shows that only about 5% of the variance in pulse rate is accounted for by height. The last section of Output 4.2 gives the estimated intercept and slope for the model. The slope is estimated to be 0.21, which implies that for every centimeter increase in height, pulse rate increases by 0.21. But because the standard error of the estimated slope is 0.14, the 95% confidence interval for the slope is [-0.07, 0.49], which includes the value 0, as we already knew it would from the result of the F-test discussed above.

The fit plot shown in Figure 4.4 consists of a scatter plot with the fitted line and 95% confidence limits for the line shown as a shaded area. (Details of how this confidence interval is constructed are given in Der and Everitt, 2013.) Also given in Figure 4.4 are the 95% confidence limits for predicted values of pulse rate (predicted values are simply the values of pulse rate found from applying the fitted model—that is, predicted pulse rate=estimated intercept +estimated slope x height.) Again, the details of how the confidence interval for predicted values is found are given in Der and Everitt (2013).

We can see that a horizontal line (one with slope 0) could easily be fitted between the confidence limits for the line.

Figure 4.4: Scatter Plot of Pulse and Height Data Showing Fitted Linear Regression and Confidence Interval for the Fit

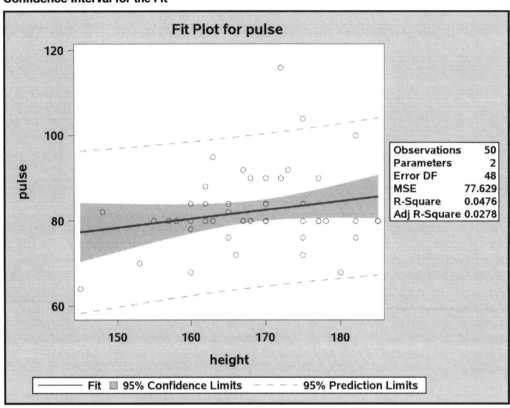

4.2.4 A Further Example of Linear Regression

Now let us consider a different set of bivariate data that were collected in an experiment in *kinesiology* (a natural health care system, which uses gentle muscle testing to evaluate many functions of the body in the structural, chemical, neurological, and biochemical realms). A subject performed a standard exercise at a gradually increasing level. Two variables were measured, the first oxygen uptake and the second expired ventilation, which is related to the rate of exchange of gases in the lungs. The data are shown in Table 4.2. Once again, the objective is to investigate the relationship between the two measured variables.

Table 4.2: Oxygen Uptake and Expired Ventilation Data

Oxygen Uptake	Expired Ventilation	Oxygen Uptake	Expired Ventilation
574	21.9	2577	46.3
592	18.6	2766	55.8

Oxygen Uptake	Expired Ventilation	Oxygen Uptake	Expired Ventilation
664	18.6	2812	54.5
667	19.1	2893	63.5
718	19.2	2957	60.3
770	16.9	3052	64.8
927	18.3	3151	69.2
947	17.2	3161	74.7
1020	19.0	3266	72.9
1096	19.0	3386	80.4
1277	18.6	3452	83.0
1323	22.8	3521	86.0
1330	24.6	3543	88.9
1599	24.9	3676	96.8
1639	29.2	3741	89.1
1787	32.0	3844	100.9
1790	27.9	3878	103.0
1794	31.0	4002	113.4
1874	30.7	4114	111.4
2049	35.4	4152	119.9
2132	36.1	4252	127.2
2160	39.1	4290	126.4
2292	42.6	4331	135.5
2312	39.9	4332	138.9
2475	46.2	4390	143.7
2489	50.9	4393	144.8
2490	46.5		

The data on oxygen uptake and expired ventilation, shown in Table 4.2, are available in the SAS data set **anaerob.** To produce a scatter plot for these data, repeat the scatter plot task, assigning **o2in** to the x axis and **airout** to the y. The result is shown in Figure 4.5, which clearly demonstrates that there is a strong relationship between oxygen uptake and expired ventilation, but that this relationship is distinctly non-linear; as oxygen uptake increases, expired ventilation accelerates, making the relationship between the two variables depart from a straight line form.

Figure 4.5: Scatter Plot of Oxygen Uptake and Expired Ventilation

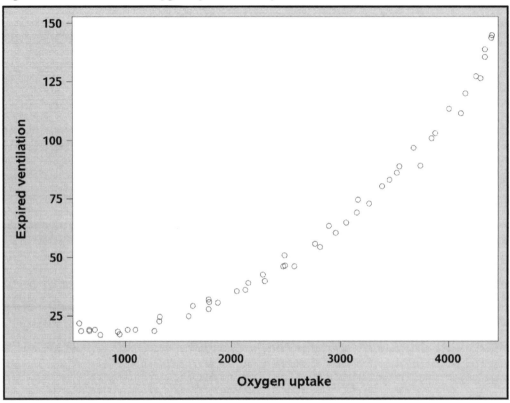

The correlation coefficient for oxygen uptake and expired ventilation can be found in the same way as described in the previous section for height and resting pulse rate. The results are shown in Output 4.3.

Output 4.3: Correlation for Oxygen Uptake and Expired Ventilation

Pearson Correlation Coefficients, N = 53 Prob > |r| under H0: Rho=0		
	o2in	airout
o2in Oxygen uptake	1.00000	0.95498 <.0001
airout Expired ventilation	0.95498 <.0001	1.00000

For oxygen uptake and expired volume, the correlation is 0.95, but here since we know the relationship to be non-linear, use of the coefficient is not totally informative about the nature of the data. So although a very small *p*-value associated with the test of zero correlation provides strong evidence that the two variables are related, the scatter plot indicates that the relationship is not linear. This example emphasizes that it is generally good practice to always have the scatter plot of two variables visible when trying to interpret the correlation coefficient between them.

The scatter plot in Figure 4.5 clearly demonstrates that the linear model is not sufficient for the expired ventilation and oxygen uptake data; such a model would not describe the data well. For these data, we need a model that includes a *quadratic effect* for oxygen uptake such as the following:

$$y_i = \alpha + \beta_1 x_i + \beta_2 x_i^2 + \varepsilon_i$$

We make the same assumptions about the error terms as in the simple linear regression model and the model is, perhaps confusingly, still an example of a *linear model* because it is linear in the parameters, α, β_1 and β_2, and can be fitted by applying linear regression. (An example of a *nonlinear* model is $y_i = \alpha_1 \exp(\beta_1 x_i) + \alpha_2 \exp(\beta_2 x_i^2) + \varepsilon_i$. We shall not deal with such models in this book.)

To fit a quadratic model to the expired ventilation data:

1. Open **Tasks ▶ Statistics ▶ Linear Regression**.
2. Under **Data ▶ Data**, enter **sasue.anearob**.
3. Under **Data ▶ Roles**, add **airout** as the **Dependent variable** and **o2in** to the **Continuous variables box**.
4. Under **Model ▶ Model Effects**, select **o2in** and click the **Add** button; then click on the button labelled **Polynomial Order=N**.
5. A popup window opens to set the value of **N**, with **2** as the default. Click **Add**. This adds the term **o2in*o2in** to the **Model Effects**.
6. Click **Run**.

The results are shown in Output 4.4. The parameter estimates part of this output show that both the linear and quadratic parts of the model are highly significant, as would be expected from looking at the scatter plot of the data.

Output 4.4: Results of Fitting a Quadratic Model to the Data in Table 4.2

Data Set	SASUE.ANAEROB
Dependent Variable	airout
Selection Method	None

Number of Observations Read	53
Number of Observations Used	53

Dimensions	
Number of Effects	3
Number of Parameters	3

Least Squares Summary			
Step	Effect Entered	Number Effects In	SBC
0	Intercept	1	393.7570
1	o2in	2	268.9281
2	o2in*o2in	3	131.6651*
* Optimal Value of Criterion			

Analysis of Variance					
Source	DF	Sum of Squares	Mean Square	F Value	Pr > F
Model	2	82340	41170	4054.90	<.0001
Error	50	507.65646	10.15313		
Corrected Total	52	82848			

Root MSE	3.18640
Dependent Mean	60.70755
R-Square	0.9939
Adj R-Sq	0.9936
AIC	180.75419
AICC	181.58752
SBC	131.66507

Parameter Estimates					
Parameter	DF	Estimate	Standard Error	t Value	Pr > |t|
Intercept	1	24.270395	1.940232	12.51	<.0001
o2in	1	-0.013441	0.001762	-7.63	<.0001
o2in*o2in	1	0.000008902	0.000000344	25.85	<.0001

4.2.5 Checking Model Assumptions: The Use of Residuals

A linear regression analysis should not end without an attempt to check the assumptions of constant variance and normality of the error terms in the model. Departures from one or other or both of these assumptions may invalidate conclusions based on the regression analysis. Checking the assumptions involves the use of one or another type of residual, the simplest of which is just the difference between an observed value of the y-variable and the value predicted by the fitted model. The residuals can be plotted in a number of ways to gain insights into whether the assumptions of constant variance and normality are reasonable:

Residuals versus fitted values: If the fitted model is appropriate, the plotted points should lie within an approximately horizontal band across the points. Departures from this appearance may indicate that the functional form of the assumed model is incorrect or, alternatively, that there is non-constant variance.

Residuals versus the x-variable: A systematic pattern in this plot can indicate departures from the constant variance assumption or an inappropriate model form.

Normal probability plot (see Der and Everitt, 2013) of the residuals: This plot is used to check the normality of the error terms in the model.

Figure 4.6 shows some idealized plots that indicate particular points about a model. Figure 4.6 (a) is what is looked for to confirm that the fitted model meets the assumptions of the regression model. Figure 4.6 (b) suggests that the assumption of constant variance is not justified, so a transformation of the *y*-variable before fitting might be a sensible option to consider. Figure 4.6 (c) implies that the model requires a quadratic term in the *x*-variable.

Figure 4.6: Idealized Residual Plots

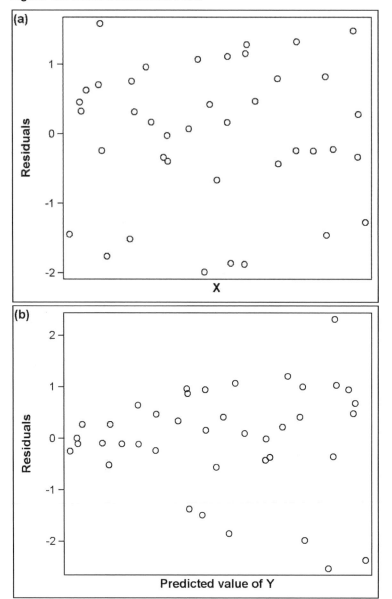

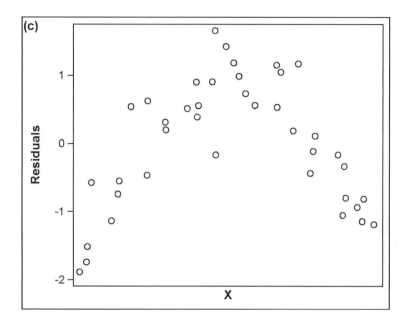

The plots we need here are produced in part of a diagnostic panel of plots produced by default by the Linear Regression task. However, when we ran the task for the height and resting pulse data, we deselected these to simplify the output. We can now produce the diagnostic panel as follows:

1. Reopen the Linear Regression task for the resting data or, if it has been closed, repeat steps 1 to 4.
2. Under **Options ▶ Plots ▶ Diagnostic and Residual Plots**, select **Diagnostic plots** and **Residuals for each explanatory variable**. For each, there is the option to display these as a **Panel of plots** (the default) or as individual plots. The panel is sufficient for most purposes, but individual plots can be helpful when larger versions are needed to show more detail.
3. Click **Run**.

The results include the diagnostic panel shown in Figure 4.7 and the residual plot shown in Figure 4.8.

Figure 4.7: Fit Diagnostics for Pulse

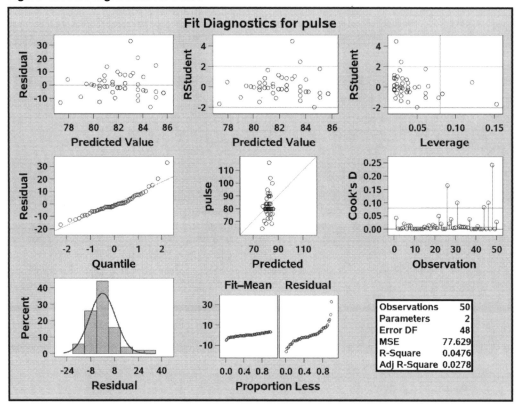

Figure 4.8: Residuals for Pulse

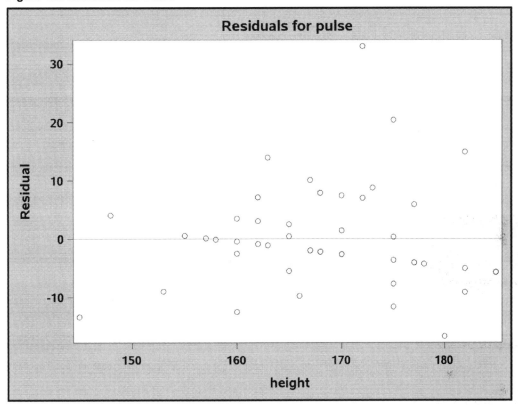

First, let us look at Figure 4.7 and concentrate on the very first plot in the panel and in the fourth and seventh plots counting across rows of the panel. (We shall return to the other plots in the diagnostic panel in Chapter 6.) The plot of residuals against predicted values shows no obvious pattern that would lead to doubting the model assumptions. The normal probability plot in panel four deviates a little for linearity but not enough to think that the normality assumption of the error terms is misleading, a point emphasized by the histogram and fitted normal distribution given in panel seven.

Moving on to Figure 4.8, we see that the residuals plotted against height show no clear pattern, although there are one or two large residuals for some of the larger height values. This might indicate outliers (observations that are considerably different from the bulk of the data and might unduly influence parameter estimates), although this seems unlikely here.

Now we can move on to the expired ventilation and oxygen uptake data set (Output 4.4), where the corresponding diagnostic panel is shown in Figure 4.9 and the other residual plots in Figure 4.10. None of the plots give any real cause for concern about the fitted model.

Figure 4.9: Diagnostic Panel for Quadratic Model Fitted to Oxygen Uptake Data

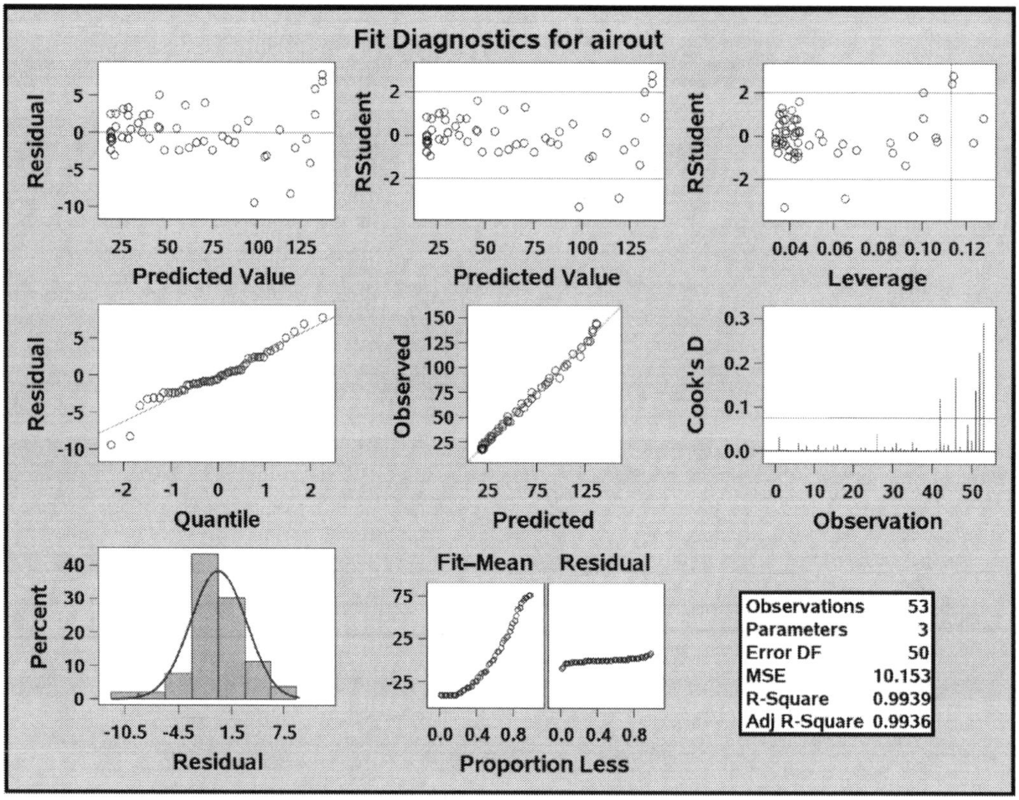

Figure 4.10: Residual by Regressors for Airout

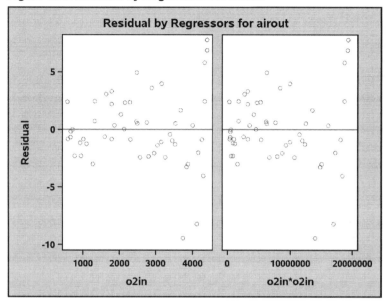

4.3 Adjusting the Scatter Plot To Show Patterns in the Data

The data in Table 4.3 (from Cook and Weisberg, 1994) give the monthly U.S. births per thousand population for the years 1940 to 1948. Here we would like to explore the data for any interesting patterns that might tell a story about the data.

Table 4.3: U.S. Monthly Birthrates Between 1940 and 1943

1890	1957	1925	1885	1896	1934	2036	2069	2060
1922	1854	1852	1952	2011	2015	1971	1883	2070
2221	2173	2105	1962	1951	1975	2092	2148	2114
2013	1986	2088	2218	2312	2462	2455	2357	2309
2398	2400	2331	2222	2156	2256	2352	2371	2356
2211	2108	2069	2123	2147	2050	1977	1993	2134
2275	2262	2194	2109	2114	2086	2089	2097	2036
1957	1953	2039	2116	2134	2142	2023	1972	1942
1931	1980	1977	1972	2017	2161	2468	2691	2890
2913	2940	2870	2911	2832	2774	2568	2574	2641
2691	2698	2701	2596	2503	2424			

(read along rows for temporal sequence)

4.3.1 Plotting the Birthrate Data: The Aspect Ratio of a Scatter Plot

An important aspect of a scatter plot that can greatly influence our ability to recognize patterns in the plot is the *aspect ratio*, the physical length of the vertical axis relative to that of the horizontal axis. To illustrate how changing the aspect ratio of a scatter plot can help us understand what the data might be trying to say, we shall use the birthrate data given in Table 4.3.

The data are in the data set **usbirths.** We can construct a scatter plot of the birthrates against the date with the default aspect ratio:

1. Open **Tasks ▶ Graph ▶ Scatter plot**.
2. Under Data ▶ Data, enter **sasue.usbirths**.
3. Under **Data ▶ Roles**, add **obsdate** as the **X variable** and **rate** as the **Y variable**.
4. Under **Options ▶ X axis**, deselect **Show grid line**. Do the same for **Y axis**.
5. Under **Options ▶ Graph Size**, the defaults can be seen to be 4.8 by 6.4 inches, which gives an aspect of 3:4.
6. Click **Run**.

The resulting plot is shown in Figure 4.11. The plot shows that the U.S. birthrate was increasing between 1940 and 1943, decreasing between 1943 and 1946, rapidly increasing during 1946, and then decreasing again between 1947 and 1948. As Cook and Weisberg comment:

These trends seem to deliver an interesting history lesson since the U.S. involvement in World War II started in 1942 and troops began returning home during the part of 1945, about nine months before the rapid increase in the birth rate.

Figure 4.11: Scatter Plot of Birthrate vs Month

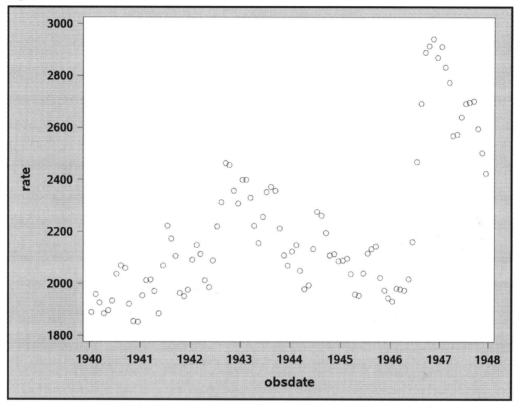

Now let us see what happens when we alter the aspect ratio of the plot:

1. Reopen the previous task, or repeat the steps above.
2. Under **Options ▶ Graph Size**, enter **6** for the width and **2** for the height.
3. Click **Run**.

The resulting graph appears in Figure 4.12.

Figure 4.12: Scatter Plot of Birthrate vs Month with Aspect Ratio=0.3

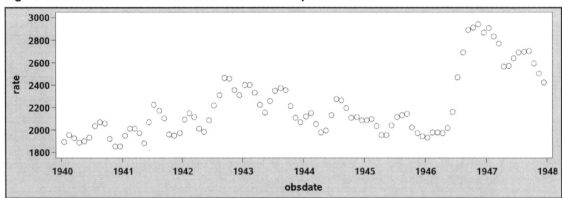

The new plot displays many peaks and troughs and suggests perhaps some minor within-year trends in addition to the global trends apparent in Figure 4.11. A clearer picture is obtained by plotting only a part of the data; here we will plot observations for the years 1940-1943. To do this, we begin by creating a filter to include only data for the years 1940-1943. We could do this by creating a filter to include only data for the years 1940-1943. However, the scatter plot task has the facility built using the WHERE clause filter.

1. Reopen the Scatter Plot task.

2. Under **Data ▶ Where Clause Filter**, select **Apply where clause** and, in the **Where string** box, enter **year<1944**.

3. Rerun the task.

The result is shown in Figure 4.13.

Figure 4.13: Scatter Plot of Birthrate vs Month for Years 1940-1943 with Aspect Ratio=0.3

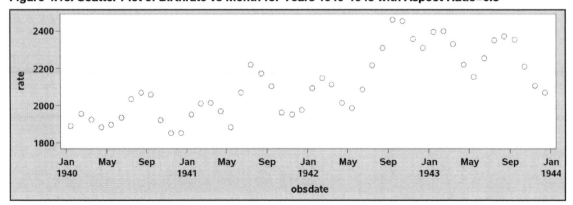

Now, a within-year cycle is clearly apparent, with the lowest within-year birthrate at the beginning of the summer and the highest occurring in the autumn. This pattern can be made clearer in a line plot with the same options as Figure 4.13:

1. Open **Tasks ▶ Graph ▶ Series Plot**.
2. Repeat the options for Figure 4.11.
3. Under **Options ▶ Graph Size**, enter **6** for the width and **2** for the height.
4. Under **Data ▶ Where Clause Filter**, select **Apply where clause** and, in the **Where string** box, enter **year<1944**
5. Click **Run**.

The new plot appears in Figure 4.14.

Figure 4.14: Scatter Plot of Birthrate vs Month for Years 1940-1943 with Observations Joined and Aspect Ratio=0.3

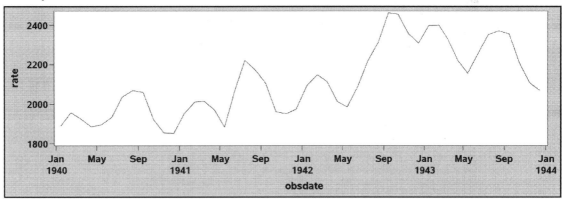

By reducing the aspect ratio to 0.25 and replotting all 96 observations with a line plot, both the within-year and global trends become clearly visible.

1. Click on the **Series Plot** tab.
2. Under **Data** deselect **Apply where clause**.
3. Under **Options** set the width to 8 and the height to 2.

The result is shown in Figure 4.15.

Figure 4.15: Scatter Plot of Birthrate vs Month with Observations Joined and Aspect Ratio=0.25

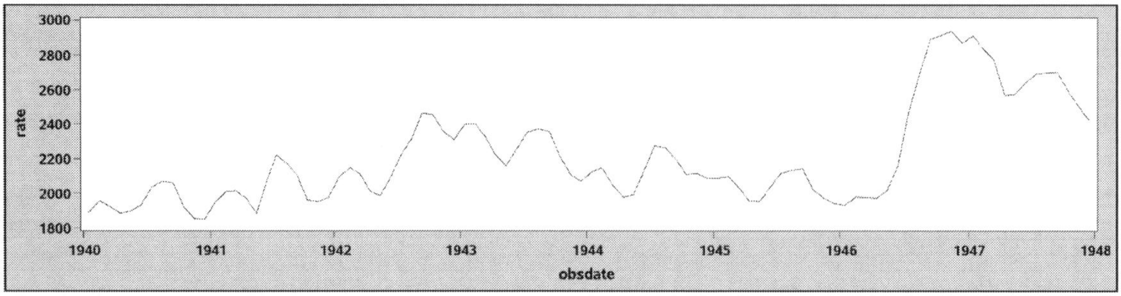

4.4 Exercises

Exercise 4.1: Mortality Data

The **mortality** data set contains mortality rates due to malignant melanoma of the skin for white males during the period 1950-1969 for the District of Columbia and each state on the US mainland. Also given are the latitude and longitude of the centre of each state. Construct scatter plots of mortality against latitude and mortality against longitude; in each case, find the corresponding correlation coefficient. Interpret your findings.

Mortality Rates Due to Malignant Melanoma in the USA

State	Mortality	Latitude	Longitude
Alabama	219	33.0	87.0
Arizona	160	34.5	112.0
Arkansas	170	35.0	92.5
California	182	37.5	119.5
Colorado	149	39.0	105.5
Connecticut	159	41.8	72.8
Delaware	200	39.0	75.5
Washington, DC	177	39.0	77.0
Florida	197	28.0	82.0
Georgia	214	33.0	83.5
Idaho	116	44.5	114.0
Illinois	124	40.0	89.5

State	Mortality	Latitude	Longitude
Iowa	128	42.2	93.8
Kansas	166	38.5	98.5
Kentucky	147	37.8	85.0
Louisiana	190	31.2	91.8
Maine	117	45.2	69.0
Maryland	162	39.0	76.5
Massachusetts	143	42.2	71.8
Michigan	117	43.5	84.5
Minnesota	116	46.0	94.5
Mississippi	207	32.8	90.0
Missouri	131	38.5	92.0
Montana	109	47.0	110.5
Nebraska	122	41.5	99.5
Nevada	191	39.0	117.0
New Hampshire	129	43.8	71.5
New Jersey	159	40.2	74.5
New Mexico	141	35.0	106.0
New York	152	43.0	75.5
North Carolina	199	35.5	79.5
North Dakota	115	47.5	100.5
Ohio	131	40.2	82.8
Oklahoma	182	35.5	97.2
Oregon	136	44.0	120.5
Pennsylvania	132	40.8	77.8
Rhode Island	137	41.8	71.5
South Carolina	178	33.8	81.0
South Dakota	86	44.8	100.0
Tennessee	186	36.0	86.2
Texas	229	31.5	98.0
Utah	142	39.5	111.5
Vermont	153	44.0	72.5
Virginia	166	37.5	78.5

State	Mortality	Latitude	Longitude
Washington	117	47.5	121.0
West Virginia	136	38.8	80.8
Wisconsin	110	44.5	90.2
Wyoming	134	43.0	107.5

Exercise 4.2: Index Data

The **index** data set gives the values of a food price index and a house price measure for the UK for each year from 1971 to 1989. By constructing suitable plots, investigate how the two price measures change over time, and how the changes are related. (Experiment with changing the aspect ratio of the plots you create.)

Year	Food Index	House Index
1971	155.6	60
1972	169.4	79
1973	194.9	107
1974	230.0	113
1975	288.9	124
1976	346.5	134
1977	412.4	148
1978	441.6	177
1979	494.7	227
1980	554.5	272
1981	601.3	280
1982	648.6	285
1983	669.2	317
1984	706.7	342
1985	728.8	373

Year	Food Index	House Index
1986	752.6	436
1987	775.6	513
1988	802.4	646
1989	847.7	750

Exercise 4.3: Alcohol and Tobacco Data (the expenditure dataset)

Moore and McCabe (1989) give the data below, taken from the Family Expenditure Survey of the British Department of Employment. (The data are also given in Howell, 2002.) The two variables are the average weekly household expenditure on alcohol and on tobacco, each in pounds.

Region	Alcohol	Tobacco
North	6.47	4.03
Yorkshire	6.13	3.76
Northeast	6.19	3.77
East Midlands	4.89	3.34
West Midlands	5.63	3.47
East Anglia	4.52	2.92
Southeast	5.89	3.20
Southwest	4.79	2.71
Wales	5.27	3.53
Scotland	6.08	4.51
Northern Ireland	4.02	4.56

1. Construct a scatter plot of the data, identifying the regions in some suitable way.
2. Calculate the correlation coefficient for the two variables.
3. Are there any regions you think should be left out of the calculation of the correlation and why? If you do recalculate the correlation with this observation (these observations), removed and compare with the result found in Step 2.

Exercise 4.4: IQ Data (the iqs dataset)

Miles and Shevlin (2001) give the data shown below, which record the IQs of 30 heterosexual couples.

Couple	Male IQ	Female IQ
1	74	75
2	126	125
3	134	141
4	96	77
5	78	91
6	66	64
7	80	66
8	92	87
9	92	87
10	92	95
11	60	138
12	98	111
13	90	89
14	118	134
15	86	109
16	68	86
17	118	122
18	98	116
19	114	107
20	116	111
21	84	71
22	92	80
23	100	100
24	144	136
25	86	71
26	128	104
27	86	77
28	110	114
29	112	118

Couple	Male IQ	Female IQ
30	78	68

1. Construct a scatter plot of the data and identify any observation that might distort any analyses of the data.
2. Find the correlation coefficient of the two IQ scores with and without any observations identified in Step 1.
3. Fit a linear regression for predicting female partner IQ from the male partner IQ and show the fit on a scatter plot.
4. Examine the residuals from the fitted model and plot them in any way you feel is appropriate. Does your plot (plots) give you any cause for concern over the fitted model?
5. What light does the data throw on the hypothesis that people seek partners with a similar IQ to their own?

Exercise 4.5: Galaxy Data (galaxies dataset)

Freedman et al. (2001) give the relative velocity and the distance of 24 galaxies according to the measurements made using the Hubble Space Telescope. The data are given below.

ID	Galaxy	Velocity	Distance
1	NGC0300	133	2.00
2	NGC0925	664	9.16
3	NGC1326A	1794	16.14
4	NGC1365	1594	17.95
5	NGC1425	1473	21.88
6	NGC2403	278	3.22
7	NGC2541	714	11.22
8	NGC2090	882	11.75
9	NGC3031	80	3.63
10	NGC3198	772	13.80
11	NGC3351	642	10.00
12	NGC3368	768	10.52
13	NGC3621	609	6.64
14	NGC4321	1433	15.21
15	NGC4414	619	17.70
16	NGC4496A	1424	14.86
17	NGC4548	1384	16.22

ID	Galaxy	Velocity	Distance
18	NGC4535	1444	15.78
19	NGC4536	1423	14.93
20	NGC4639	1403	21.98
21	NGC4725	1103	12.36
22	IC4182	318	4.49
23	NGC5253	232	3.15
24	NGC7331	999	14.72

1. Construct a scatter plot of the data.
2. Fit a *suitable* linear model to the data.
3. Use the fitted model to estimate the age of the universe (see Der and Everitt, 2014).

Exercise 4.6: Race Data

The data below show the times of the winners of the 1500 m race for men in each Olympic Games from 1896 until 2004. The dataset is **olympic1500** and the time is in seconds.

Year	Venue	Winner	Country	Time (minutes and seconds)
1896	Athens	P Flack	Australia	4-33.2
1900	Paris	C. Bennett	Great Britain	4-06.2
1904	St Louis	J. Lightbody	USA	4-05.4
1908	London	M.Sheppard	USA	4-03.4
1912	Stockholm	A. Jackson	Great Britain	3-56.8
1920	Antwerp	A.Hill	Great Britain	4-01.8
1924	Paris	P. Nurmi	Finland	3-53.6
1928	Amsterdam	H. Larva	Finland	3-53.2
1932	Los Angeles	L. Beccali	Italy	3-51.2
1936	Berlin	J. Lovelock	New Zealand	3-47.8
1948	London	H. Eriksson	Sweden	3-49.8
1952	Helsinki	J. Barthel	Luxemborg	3-45.1
1956	Melbourne	R. Delaney	Ireland	3-41.2
1960	Rome	H. Elliot	Australia	3-35.6
1964	Tokyo	P. Snell	New Zealand	3-38.1

Year	Venue	Winner	Country	Time (minutes and seconds)
1968	Mexico City	K. Keino	Kenya	3-34.9
1972	Munich	P. Vasala	Finland	3-36.3
1976	Montreal	J. Walker	New Zealand	3-39.2
1980	Moscow	S. Coe	Great Britain	3-38.4
1984	Los Angeles	S. Coe	Great Britain	3-32.5
1988	Seoul	P. Rono	Kenya	3-35.9
1992	Barcelona	F. Cacho	Spain	3-40.1
1996	Atlanta	N. Morceli	Algeria	3-35.8
2000	Sydney	K. Ngenyi	Kenya	3-32.1
2004	Athens	H. El Guerrouj	Morocco	3-34.2

1. Construct a scatter plot of the data. Are there any observations that you think should be removed before further analysis?

2. Fit a linear regression to the data (after removing any observations identified from the scatter plot) and show the line on the scatter plot of the data used.

3. Do you think the linear model is a sensible model for these data? Give your reasons.

4. If you think the linear model is inadequate for these data, fit what you think might be a preferable model and plot it on the scatter plot of the data used.

5. Use your model to construct 95% confidence intervals for the predicted times of the winners of the men's 1500 m in both the 2008 Bejing Olympics and in the 2012 London Olympics.

Chapter 5: Analysis of Variance

5.1 Introduction

In this chapter, we shall describe how to analyse data in which a response variable of interest is measured in different levels of one or more categorical *factor variables*. The statistical topics to be covered are:

- Analysis of variance for the one-way design
- Factorial designs, balanced and unbalanced
- Type I and Type III sums of squares
- Multiple comparison tests

5.2 One-Way ANOVA

In an experiment to compare different methods of teaching arithmetic (Wetherill, 1982), 45 students were divided randomly into five groups of equal size. Two groups, 1 and 2, were taught by the current method and three, 3-praised, 4-reproved, and 5-ignored, by one of three new methods. At the end of the investigation, all pupils took a standard test with the results shown in Table 5.1. What conclusions can be drawn about possible differences between teaching methods?

Table 5.1: Data on Teaching Methods

Teaching Method	Test Results
1	17,14,24,20,24,23,16,15,24
2	21,23,13,19,13,19,20,21,16
3	28,30,29,24,27,30,28,28,23
4	19,28,26,26,19,24,24,23,22
5	21,14,13,19,15,15,10,18,20

Before beginning any formal analysis of the data sets in the previous section, it will be useful to consider some summary statistics and graphics for the data since both will be helpful in gaining informal insights into the data and aiding the interpretation of the formal testing to be described later.

5.2.1 Initial Examination of Teaching Arithmetic Data with Summary Statistics and Box Plots

The teaching data in Table 5.1 are available in the SAS data set **teaching**.

We saw in Chapter 3 how to produce tables of counts and percentages. Here, we need tables containing other summary statistics, specifically means and standard deviations. To do this:

1. Open **Tasks ▶ Statistics ▶ Summary Statistics**.
2. Under **Data ▶ Data**, add **sasue.teaching**.
3. Under **Data ▶ Roles, Classification variables** are those whose values will be used to form the rows and/or columns of the table and **Analysis variables** are those whose values will be summarized within the table. In this case, **Method** is a classification variable and **Result** an analysis variable.
4. Under **Options ▶ Basic Statistics**, select **Mean** and **Standard deviation**.
5. Click **Run**.

The results are shown in Output 5.1.

Output 5.1: Summary Statistics for Teaching Methods Data

		Analysis Variable : result	
method	N Obs	Mean	Std Dev
1	9	19.6666667	4.2130749
2	9	18.3333333	3.5707142
3	9	27.4444444	2.4551533
4	9	23.4444444	3.0867099
5	9	16.1111111	3.6209268

While this provides the information we need, the large number of significant digits displayed is both unnecessary and distracting. There is no setting within the task to limit the number of decimal places, but it can be easily achieved by a small edit to the code the task generates. In the code pane, click the **Edit** button and at the end of the PROC MEANS statement before the semicolon, add `maxdec=2`.

A useful graphic for these data consists of the box plots of the observations made under each teaching method. This is available as an option within the task by selecting **Options ▶ Plots ▶ Comparative Box Plot** and rerunning the task. (An equivalent plot could also be produced by selecting **Tasks ▶ Graph ▶ Box Plot**.) The resulting box plots are shown in Figure 5.1.

The box plot and the summary statistics in Output 5.1 suggest some interesting differences between the five methods. Method 3 (praised), for example, appears to give far better results than the others, although there are two distinct outliers that perform considerably less well than the other students taught by method 3. The observations for teaching method five are quite skewed.

Figure 5.1: Box Plots from the Results under Five Teaching Methods

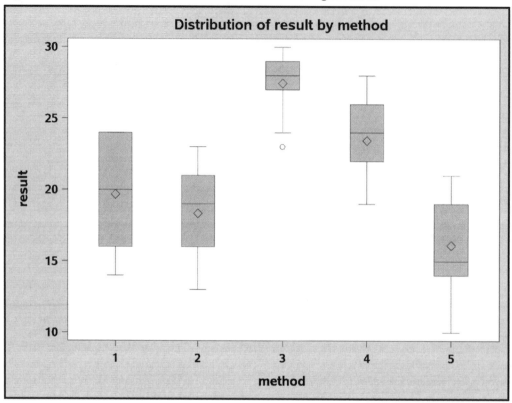

5.2.2 Teaching Arithmetic: Are Some Methods for Teaching Arithmetic Better than Others?

The teaching arithmetic study is an example of what is generally known as a *one-way design*; in such designs, interest centres on assessing the effect of a single factor variable (teaching method here) on a response variable (test score). The question posed in such a design is 'Do the populations corresponding to the different levels of the factor variable have different means?' Consequently, the null hypothesis that we aim to test is the equality of means of the populations:

$$H_0 : \mu_1 = \mu_2 = \ldots\ldots = \mu_k$$

where $\mu_1, \mu_2, \ldots \mu_k$ are the population means and k is the number of levels of the factor variable.

In Chapter 2, we described how to use Student's *t*-test to test the equality of *two* population means and here we might pose the question: 'Why not simply apply the test to each pair of means in our one-way design to assess the null hypothesis above?' The reason that such an approach is not appropriate is that if each of the $N=k(k-1)/2$ t-tests we would perform is tested at the usual 5% significance level, the probability of rejecting the equality of at least one pair of population means when the null hypothesis is true (P) is greater than the nominal significance level of 0.05 when k is three or larger. Details of the calculations involved that demonstrate that this is so are given in Everitt (1996). Here we simply give some numerical results that illustrate the problem with the t-test approach:

k	N	P
3	3	0.14
4	6	0.26
10	45	0.90

The appropriate approach to the analysis of data arising from a one-way design is the *analysis of variance (ANOVA),* the phrase having been coined by Ronald Aylmer Fisher, who defined it as the separation of variance ascribable to one group of causes from the variance ascribable to the other groups. Stated another way, the ANOVA is a partitioning of the total variance in a set of data into a number of component parts; in a one-way design, for example, we separate the total variance into a part due to differences in the sample means of the levels of the factor variable (*between-groups variance*) and a part measuring variance within the levels of the factor variable (*within-groups variance*).

If the null hypothesis of equality of means is correct, then both the between-groups variance and the within-groups variance are estimating the same population quantity; if the null hypothesis is wrong, then the between-groups variance is estimating a *larger* population quantity than the within-groups variance. Consequently, a test of the equality of the two population variances (between groups and within groups) based on the two estimates of them will be a test of the null hypothesis about the population means that we are interested in. The appropriate test for the equality of two variances is what is known as an *F-test*. (Full details of the analysis of variance for a one-way design are given in Everitt, 1996.)

The data collected in a one-way design need to satisfy the following assumptions to make the F-test involved strictly valid:

- The observations in each level of the factor variable arise from a population with a normal distribution.
- The population variances of the different levels of the factor are the same.
- The observations are independent of one another.

The assumptions are often difficult to check, particularly when the number of observations in each group is small; tests of the normality and equal variances *are* available, but they are rarely taken very seriously because they themselves are based on assumptions that are difficult to investigate.

Fortunately, the F-test is known to be relatively *robust* against departures from both normality and homogeneity of variance, especially when the number of observations in each group is equal or approximately equal. (In some cases, a *transformation* of the data, for example, taking logs, might aid in achieving both a normal distribution and homogeneity, although interpretation could become more problematical; see Exercise 5.3 for an example of applying a transformation and Everitt, 1996, for more details on transformations.)

To apply the analysis of variance to the teaching method data in Table 5.1:

1. Open **Tasks ▶ Statistics ▶ One-Way ANOVA**.
2. Under **Data ▶ Data**, add **sasue.teaching** (if necessary).
3. Under **Data ▶ Roles**, make **result** the **Dependent variable** and **method** the **Categorical variable**.
4. Click **Run**.

The results are shown in Output 5.3; in this output, it is the part that gives the result of the partition of the variation in the data that is of most importance for us because this is where we find the result of the F-test for assessing the equality of means hypothesis. We see that the *p*-value associated with the F-test is very small (<0.001), so there is considerable evidence that the teaching methods do indeed differ in respect to their mean arithmetic test scores. Levene's test for homogeneity of variance (see Everitt and Skrondal, 2010, for a definition of the test) is non-significant, which is reassuring.

The least squares means are adjusted marginal means (that is, what the model predicts the means would be if the number of observations in each group was equal). In our example, the number of subjects in each group *is* equal, so here the means and least square means are equal.

Having shown that there is strong evidence of the effect of teaching method on test scores, we might wish to investigate in more detail which methods differ (an overall significant F-test does *not* imply that *all* means differ). For the more detailed examination, we can use one of a variety of what are termed *multiple comparison tests*. Such tests compare each pair of means in turn but take steps to avoid the problem of inflating the type I error discussed earlier in the chapter.

The default multiple comparison procedure is the Tukey test, which is described in detail in Everitt (1996). The results of the Tukey test for each pair of teaching methods are given in Output 5.3, just before Figure 5.2 (of which more later). Examining the *p*-values, we see that method 1 differs from method 3, method 2 differs from methods 3 and 4, method 3 differs from method 5, and method 4 differs from method 5. So the Tukey tests produce a grouping of the teaching methods into (3,4),(1,2,5).

The plot following the Tukey test, is known as a *mean-mean scatter plot* but referred to in SAS as a *diffogram*, needs some explanation. The means of each group are arranged along both the *x* and *y* axes. Each of the 45 degree lines corresponds to a pairwise comparison--namely the pair whose means intersect at the middle of the line. If a line crosses the dotted diagonal line, the pairwise difference is non-significant; significant and non-significant differences are also indicated by different line attributes (pattern and/or colour).

Output 5.3: Edited Output for the Analysis of Variance Results for Teaching Method Data

Class Level Information		
Class	**Levels**	**Values**
method	5	1 2 3 4 5

Number of Observations Read	45
Number of Observations Used	45

Source	DF	Sum of Squares	Mean Square	F Value	Pr > F
Model	4	722.666667	180.666667	15.27	<.0001
Error	40	473.333333	11.833333		
Corrected Total	44	1196.000000			

Levene's Test for Homogeneity of result Variance ANOVA of Squared Deviations from Group Means					
Source	DF	Sum of Squares	Mean Square	F Value	Pr > F
method	4	544.0	136.0	1.38	0.2583
Error	40	3943.6	98.5901		

Level of method	N	result	
		Mean	**Std Dev**
1	9	19.6666667	4.21307489
2	9	18.3333333	3.57071421
3	9	27.4444444	2.45515331

Level of method	N	result	
		Mean	Std Dev
4	9	23.4444444	3.08670986
5	9	16.1111111	3.62092683

method	result LSMEAN	LSMEAN Number
1	19.6666667	1
2	18.3333333	2
3	27.4444444	3
4	23.4444444	4
5	16.1111111	5

Least Squares Means for effect method Pr > |t| for H0: LSMean(i)=LSMean(j) Dependent Variable: result					
i/j	1	2	3	4	5
1		0.9222	0.0002	0.1568	0.2034
2	0.9222		<.0001	0.0242	0.6496
3	0.0002	<.0001		0.1189	<.0001
4	0.1568	0.0242	0.1189		0.0005
5	0.2034	0.6496	<.0001	0.0005	

Figure 5.2: Diffogram for Teaching Method Data

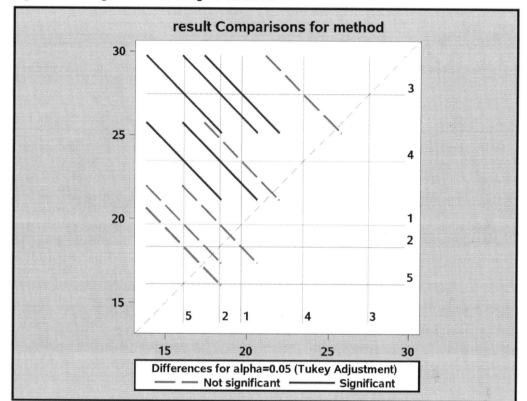

5.3 Two-Way ANOVA

The data shown in Table 5.2 come from an experiment to study the gain in weight of rats fed four different diets, distinguished by amount of protein (low and high) and by source of protein (beef and cereal). Ten rats were randomized to each of the four possible diets. The question of interest is how diet affects weight gain.

Table 5.2: Rat Weight Gain for Diets Differing by the Amount of Protein and the Source of Protein

Beef		Cereal	
Low	High	Low	high
90	73	107	98
76	102	95	74
90	118	97	56
64	104	80	111
86	81	98	95
51	107	74	88
72	100	74	82
90	87	67	77
95	117	89	86
78	111	58	92

5.3.1 A First Look at the Rat Weight Gain Data Using Box Plots and Numerical Summaries

The data on weight gain in rats given in Table 5.2 are in the data set **sasue.weight**. The data set contains three variables: **weightgain, source**, and **level**:

1. Open **Tasks ▶ Graph ▶ Box Plot**.
2. Under **Data ▶ Data**, enter **sasue.weight** in the **data** box.
3. Under **Data ▶ Roles**, add **weightgain** as the **Analysis variable**, **source** as the **Category variable**, and **level** as the **Group level**.
4. Click **Run**.

The result is shown in Figure 5.3.

Figure 5.3: Box Plots for Weight Gain in Rats Data

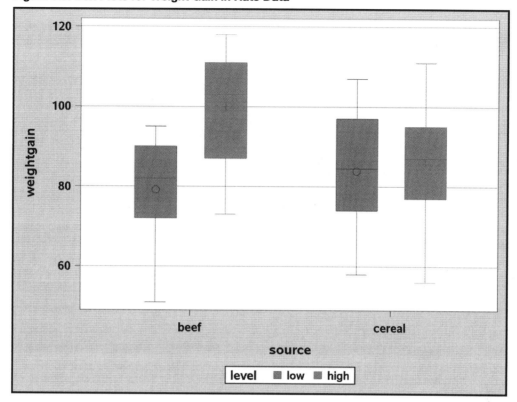

Again, it is also useful to have some numerical summaries for these data and here a 2 x 2 table formed from the two levels of source of protein and the two levels of amount of protein showing the corresponding mean and standard deviation is useful. On this occasion, we will edit the code to limit the output to two decimal places:

1. Open **Tasks ▶ Statistics ▶ Summary Statistics**.
2. Under **Data ▶ Data**, add **sasue.weight**.
3. Under **Data ▶ Roles**, assign **source** and **level** as **Classification variables** and **weightgain** as the **Analysis variable**.
4. Under **Options ▶ Basic Statistics**, select **Mean** and **Standard deviation** and deselect the others.
5. In the code pane, click the **Edit** button. A new program window opens that contains the following code:

```
proc means data=SASUE.WEIGHT chartype mean std n vardef=df;
    var weightgain;
    class source level;
```

```
run;
```

6. On the PROC MEANS statement before the semicolon, type **maxdec=2** so that it ends **vardef=df maxdec=2;**. Take care to include the semicolon at the end.

7. Click **Run**.

The results are shown in Output 5.4. The standard deviations in each cell are seen to be very similar to each other, a finding that has implications for the formal analysis of the data (see below). The mean weight gain for beef/high is considerably larger than the other three means, which are quite close to each other.

Output 5.4: Numerical Summary Statistics for Rat Weight Gain Data

Analysis Variable : weightgain					
source	level	N Obs	Mean	Std Dev	N
beef	high	10	100.00	15.14	10
	low	10	79.20	13.89	10
cereal	high	10	85.90	15.02	10
	low	10	83.90	15.71	10

5.3.2 Weight Gain in Rats: Do Rats Gain More Weight on a Particular Diet?

The data in Output 5.1 are from a simple example of what is known as a *factorial design*, which involves the simultaneous study of the effect of two or more factor variables on a response variable of interest; in the rats example, the two factors are the source and the amount of protein given to the rats. As in the previous example involving different teaching methods, questions of interest about these data concern the equality of weight gain for the two levels of source of protein and for the two levels of amount of protein. So why not apply a one-way analysis of variance (in our particular example, there are only two levels, so a one-way analysis of variance is equivalent to the *t*-test covered in Chapter 2) to each factor separately?

The answer to this question is that such an approach would omit an aspect of a factorial design that is often very important--namely, testing whether there is an *interaction* between the two factors. In simple terms, such an effect arises when the effect of applying both factors is either larger (or smaller) than the sum of the effects associated with applying each factor separately. The analysis of variance for a factorial design will include tests for such possible interaction effects. (See Everitt, 1996, for full details of the analysis of variance for factorial designs.)

To apply the analysis of variance to the data in the rat weight gain data, use the N-Way ANOVA task:

1. Open **Tasks ▶ Statistics ▶ N-Way ANOVA**.
2. Under **Data ▶ Data**, add **sasue.weight**.
3. Under **Data ▶ Roles**, assign **weightgain** as the **Dependent variable** and **source** and **level** as **Classification variables**.
4. Under **Model**, to set up a factorial model, including both main effects and their interaction, select both **source** and **level** (Ctrl-Click on both). The **Single Effects** and **Standard Models** buttons all become active. The **Add** and **Cross** buttons could be used to include the main effects and interaction (**Cross**) in the model separately, but the **Full Factorial** button does both. Clicking on that results in **source**, **level**, and **source*level** being inserted in the model effects pane.
5. Under **Options ▶ Plots ▶ Select plots to display**, we will choose **suppress plots** for the time being.
6. Click **Run**.

The results from running that model are shown in Output 5.5.

Output 5.5: Edited Output for ANOVA of Weight Gain in Rats Data

Class Level Information		
Class	**Levels**	**Values**
source	2	beef cereal
level	2	high low

Source	DF	Sum of Squares	Mean Square	F Value	Pr > F
Model	3	2404.10000	801.36667	3.58	0.0230
Error	36	8049.40000	223.59444		
Corrected Total	39	10453.50000			

Source	DF	Type I SS	Mean Square	F Value	Pr > F
source	1	220.900000	220.900000	0.99	0.3269
level	1	1299.600000	1299.600000	5.81	0.0211
source*level	1	883.600000	883.600000	3.95	0.0545

Source	DF	Type III SS	Mean Square	F Value	Pr > F
source	1	220.900000	220.900000	0.99	0.3269
level	1	1299.600000	1299.600000	5.81	0.0211
source*level	1	883.600000	883.600000	3.95	0.0545

The analysis of variance table in Output 5.5 shows the results of partitioning the variation in the weight gain observations into parts due to the amount of protein, the source of protein, and the interaction of amount and source. The corresponding F-tests show that there is evidence of a difference in weight gain for low and high levels of protein, but no evidence of a difference for source of protein. The F-test for the interaction of the two factors just fails to reach significance at the conventional 5% level, but it might still be of interest to examine in more detail just what such an interaction, if it existed, implies. The test for interaction assesses whether the difference between the mean weight gain for, say, beef and cereal protein when given at the low level is the same as the corresponding difference when given at the high level. Here there is some relatively weak evidence that the two putative differences might not be the case. To see more clearly what is happening, we can construct what is sometimes called an *interaction plot*.

An interaction plot is essentially a plot of the four cell means and is an option available within the N-Way ANOVA task.

1. Reopen **N-Way ANOVA**.
2. Under **Options ▶ Plots ▶ Select plots to display**, we now choose **Selected plots**.
   Four types of plot appear, including the **Interaction plot**, which we select. In fact, this is included among the default plots when the model includes an interaction.
3. Click **Run**.

The result is shown in Figure 5.4. The plot suggests that the difference in weight gain for beef and cereal protein is greater when given at the high level than when given at the low level, although the effect does not reach the conventional 5% significance level.

Figure 5.4: Interaction Plot for Rat Weight Gain Data

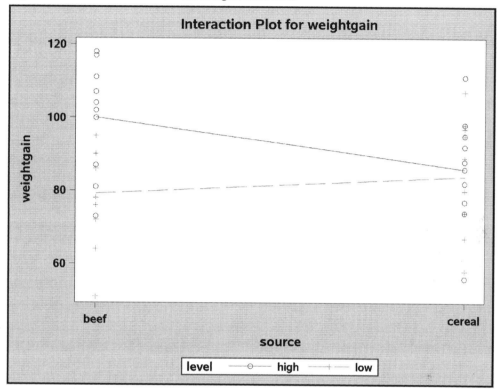

In Output 5.5, there are two analysis of variance tables, which are identical except in the labelling of the sums of squares (SS) column where one is labelled *Type I SS* and the other *Type III SS*. In the rat weight gain example where there are an equal number of observations in each cell of the factorial design, the two ways of computing sums of squares are the same, but in an unbalanced design where the number of observations differ, the Type I and Type III sums of squares are *not* the same as we will illustrate in the next section.

In a factorial design, the assumptions needed for the F-tests in the analysis of variance table to be strictly valid are similar to the assumptions needed for the one-way design listed earlier--namely, normality and homogeneity. The homogeneity assumption at least seems appropriate for the rat weight gain data given that we found in Output 5.4 that the standard deviations of the observations in each of the four cells of the design are approximately equal.

5.4 Unbalanced ANOVA

The data shown in Table 5.3 were obtained from an investigation into the effect of a mother's post-natal depression on child development. Mothers who gave birth to their first-born child in a major teaching hospital in London were divided into two groups, depressed or not depressed, on the basis of their mental state 3 months after the birth. The children's fathers were also divided into two groups: fathers who had a history of psychiatric illness and fathers who did not. The dependent variable in the study was the child's IQ at age 4 years. Only girl babies were involved in the study. The question of interest here is how post-natal depression affects a child's cognitive development.

Table 5.3: Data Obtained in a Study of the Effect of Post-Natal Depression of the Mother on the Child's Cognitive Development

Mother's Depression	Father's History	Child's IQ at 4 Years
1	0	103
1	0	124
1	0	124
1	0	104
1	0	96
2	0	92
1	0	125
1	0	99
1	0	103
1	1	98
1	1	101
1	0	104
2	1	97
2	1	95
1	0	120
2	0	105
1	0	124
1	0	123
1	0	115
1	1	110
1	1	112
1	1	120
1	1	123
2	0	98
1	0	104

Mother's Depression	Father's History	Child's IQ at 4 Years
2	0	97
1	0	125
1	0	123
1	1	101
2	1	99
1	0	120
2	0	101
1	0	118
1	0	120
1	1	99

Mother's depression: 1=no, 2=yes

Father's history: 0=no previous psychiatric history, 1=previous psychiatric history

5.4.1 Summarizing the Post-Natal Depression Data

The summary statistics and the count of the number of observations in each cell of the design for the post-natal depression data are obtained in the same way as in the previous examples.

The data are found in **sasue.depressionIQ**. For the summary table:

1. Open **Tasks ▶ Statistics ▶ Summary Statistics**.
2. Under **Data ▶ Data**, add **sasue. depressionIQ**.
3. Under **Data ▶ Roles**, assign **mo_depression** and **pa_history** as **Classification variables** and **ChildIQ** as the **Analysis variable**.
4. Under **Options ▶ Basic Statistics**, select **Mean** and **Standard deviation** and **Number of observations**.
5. Click **Run**.

The results are shown in Output 5.6.

Output 5.6: Summary Statistics for IQ and Depression Data

			Analysis Variable : ChildIQ		
Mo_depression	Pa_history	N Obs	Mean	Std Dev	N
1	0	19	114.4210526	10.3188920	19
	1	8	108.0000000	9.7687549	8
2	0	5	98.6000000	4.8270074	5
	1	3	97.0000000	2.0000000	3

The statistics given in Output 5.6 suggest that average IQ is less for children whose mother suffered from post-natal depression and for children whose father had a previous psychiatric history. Notice also that the numbers of observations in each of the four cells of the table are *not* the same; the design is said to be *unbalanced*.

5.4.2 Post-Natal Depression: Is a Child's IQ Affected?

The data on post-natal depression in a mother and its effect on a child's IQ IQ have a very similar structure to the data on weight gain in rats--both data sets involve two factor variables and a response variable. But the numbers of observations in each cell of the post-natal depression data are not equal as they are for the weight gain data. The unequal cell size in the post-natal depression data has serious implications for their analysis as we will see later.

But finding the analysis of variance table for the IQ scores from the post-natal depression study is straightforward:

1. Open **Tasks ▶ Statistics ▶ N-Way ANOVA**.
2. Under **Data ▶ Data**, add **sasue.depressionIQ**.
3. Under **Data ▶ Roles**, assign **childIQ** as the **Dependent variable** and **Pa_history** and **Mo_depression**, *in that order*, as **Classification variables**.
4. Under **Model ▶ Model Effects**, to set up a model including both main effects and their interaction, select both **Pa_history** and **Mo_depression** (Ctrl-Click on both) and click **Full Factorial**. The model pane should look like Display 5.1.
5. Under **Options ▶ Plots**, choose **Selected plots** and select the **Interaction plot**.
6. Click **Run**.

The results are shown in Output 5.7.

Display 5.1: Model Pane for Post-Natal Depression Data

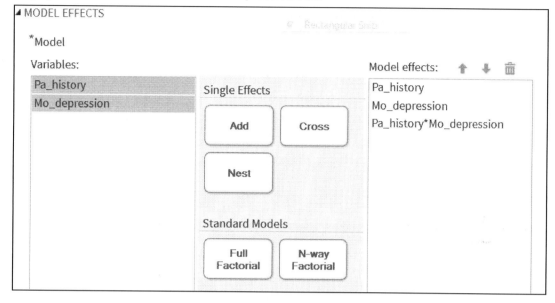

Output 5.7: Analysis of Variance Results for the Post-Natal Depression Study

Class Level Information		
Class	**Levels**	**Values**
Pa_history	2	0 1
Mo_depression	2	1 2

Number of Observations Read	35
Number of Observations Used	35

Source	DF	Sum of Squares	Mean Square	F Value	Pr > F
Model	3	1537.768421	512.589474	5.92	0.0026
Error	31	2685.831579	86.639728		
Corrected Total	34	4223.600000			

R-Square	Coeff Var	Root MSE	ChildIQ Mean
0.364090	8.523852	9.308046	109.2000

Source	DF	Type I SS	Mean Square	F Value	Pr > F
Pa_history	1	282.975000	282.975000	3.27	0.0804
Mo_depression	1	1222.101866	1222.101866	14.11	0.0007
Pa_histor*Mo_depress	1	32.691555	32.691555	0.38	0.5435

Source	DF	Type III SS	Mean Square	F Value	Pr > F
Pa_history	1	90.492912	90.492912	1.04	0.3147
Mo_depression	1	1011.820488	1011.820488	11.68	0.0018
Pa_histor*Mo_depress	1	32.691555	32.691555	0.38	0.5435

Pa_history	ChildIQ LSMEAN	H0:LSMean1=LSMean2 Pr > |t|
0	106.510526	0.3147
1	102.500000	

Pa_history	ChildIQ LSMEAN	95% Confidence Limits	
0	106.510526	101.739641	111.281411
1	102.500000	96.073924	108.926076

Least Squares Means for Effect Pa_history			
i	j	Difference Between Means	Simultaneous 95% Confidence Limits for LSMean(i)-LSMean(j)
1	2	4.010526	-3.992871 12.013924

Mo_depression	ChildIQ LSMEAN	H0:LSMean1=LSMean2 Pr > |t|
1	111.210526	0.0018
2	97.800000	

Mo_depression	ChildIQ LSMEAN	95% Confidence Limits	
1	111.210526	107.210018	115.211035
2	97.800000	90.868065	104.731935

Least Squares Means for Effect Mo_depression			
i	j	Difference Between Means	Simultaneous 95% Confidence Limits for LSMean(i)-LSMean(j)
1	2	13.410526	5.407129 21.413924

In Output 5.7, we see that the Type I and Type III sums of squares for the main effects of the father's psychiatric history and the mother's post-natal depression are *not* the same, but they *are* the same for the interaction term. In a factorial design where there are unequal numbers of observations in each cell of the design, it is no longer possible to partition the variation in the data into *non-overlapping* or *orthogonal* sums of squares representing main effects and interactions as it is in a design with equal numbers of observations in each cell.

In an unbalanced two-way layout, for example, with factors A and B, there is a proportion of the variance of the response variable that can be attributed to either A or B. The consequence is that A and B together explain less of the variation of the dependent variable than the sum of which each explains alone. The result is that the sum of squares corresponding to a factor depends on which other terms are currently in the model for the observations, so the sums of squares depend on the order in which the factors are considered and represent a comparison of models. For example, for the order A, B, A x B, the sums of squares are such that

- SSA (sum of squares for A) compares the model containing only the A main effect with one containing only the overall mean.
- SSB|A (sum of squares for B given A is already in the model) compares the model with both main effects, but no interaction, with one including only the main effect of A.
- SSAB|A,B (sum of squares for A x B given that A and B are already in the model) compares the model including an interaction and main effects with one including only main effects.

These are what are called *Type I* sums of squares. In contrast, *Type III* sums of squares represent the contribution of each term to a model including all other possible terms. Thus, for two-factor designs, the Type III sums of squares represent:

- SSA: SSA|B,AB (sum of squares for A given that B and A x B are in the model)
- SSB: SSB|A,AB (sum of squares for B given that A and A x B are in the model)
- SSAB: SSAB|A,B (sum of squares for A x B given that A and B are in the model)

(SAS also has a Type IV sum of squares, which is the same as Type III unless the design contains empty cells.)

In a balanced design, Type I and Type III sums of squares are equal, but for an unbalanced design, they are not. There have been numerous discussions over which type is most appropriate for the analysis of such designs. Authors such as Maxwell and Delaney (1990) and Howell (2002) strongly recommend the use of Type III sums of squares. Nelder (1977) and Aitkin (1978), however, are strongly critical of correcting main effects sums of squares for an interaction term involving the corresponding main effect; their criticisms are based on both theoretical and pragmatic grounds. The arguments are relatively subtle but in essence go something like the following:

When fitting models to data, the principle of parsimony is of critical importance. In choosing among possible models, we do not adopt complex models for which there is no empirical evidence.

So, if there is no convincing evidence of an AB interaction, we do not retain the term in the model. Thus, additivity of A and B is assumed unless there is convincing evidence to the contrary.

So the argument proceeds that Type III sum of squares for A in which it is adjusted for AB makes no sense.

First, if the interaction term is necessary in the model, then the experimenter will usually wish to consider simple effects of A at each level of B separately. A test of the hypothesis of no A main effect would not usually be carried out if the AB interaction is significant.

If the AB interaction is not significant, then adjusting for it is of no interest and causes a substantial loss of power in testing the A and B main effects.

(The issue does not arise so clearly in the balanced case, for there the sum of squares for a main effect is independent of whether interaction is assumed or not. Thus, in deciding on possible models for the data, the interaction term is not included unless it has been shown to be necessary, in which case tests on main effects involved in the interaction are not carried out or, if carried out, not interpreted.)

The arguments of Nelder and Aitkin against the use of Type III sums of squares are powerful and persuasive. Their recommendation--to use Type I sums of squares combined with doing a number of analyses in which main effects are considered in a number of orders as the most suitable way in which to identify a suitable model for a data set--is also convincing and strongly endorsed by the authors of this book.

So, for the post-natal depression data, we will now produce another analysis of variance table in which the main effects are considered in the order of mother's depression followed by father's psychiatric history, the reverse order to that which produced Table 5.7. To do this:

1.   Reopen **N-Way ANOVA**.

2. Under **Model ▶ Model Effects**, select **Pa_history** and click the down arrow (⬇, **Move effect down**).
3. Click **Run**.

The results are shown in Output 5.8. Comparing the analysis of variance table in Output 5.8 with that in Output 5.7, we see that the interaction Type I sums of squares are the same, but that the main effects sums of squares are different for the two ways of ordering the effects. (The Type III sums of squares are, of course, the same in both Output 5.7 and 5.8.) The conclusion to be made from both analyses is that there is no evidence of an interaction effect and no evidence that the father's psychiatric history affects the child's IQ. But there is strong evidence that the child's IQ is associated with the mother's post-natal depression, with the occurrence of post-natal depression in the mother appearing to lead to a lower IQ for the child at age 4.

Output 5.8: Analysis of Variance Results for the Post-Natal Depression Study after Reordering of Main Effects

Source	DF	Type I SS	Mean Square	F Value	Pr > F
Mo_depression	1	1300.859259	1300.859259	15.01	0.0005
Pa_history	1	204.217607	204.217607	2.36	0.1349
Mo_depres*Pa_history	1	32.691555	32.691555	0.38	0.5435

Source	DF	Type III SS	Mean Square	F Value	Pr > F
Mo_depression	1	1011.820488	1011.820488	11.68	0.0018
Pa_history	1	90.492912	90.492912	1.04	0.3147
Mo_depres*Pa_history	1	32.691555	32.691555	0.38	0.5435

5.5 Exercises

Exercise 5.1: Rats Data

The **rats** data set derives from a study in which the effects of three different poisons and four different treatments on the survival times of rats in hours were of interest. Carry out an appropriate analysis of variance of these data, paying particular attention to possible violations of distributional assumptions.

Poison	Treatment			
	A	B	C	D
1	0.31	0.82	0.43	0.45
	0.45	1.10	0.45	0.71
	0.46	0.88	0.63	0.66
	0.43	0.72	0.76	0.62
2	0.36	0.92	0.44	0.56
	0.29	0.61	0.35	1.02
	0.40	0.49	0.31	0.71
	0.23	1.24	0.40	0.38
3	0.22	0.30	0.23	0.30
	0.21	0.37	0.25	0.36
	0.18	0.38	0.24	0.31
	0.23	0.29	0.22	0.22

Exercise 5.2: Knee Data

The **knee** data set comes from an investigation described by Kapor (1981) in which the effect of knee-joint angle on the efficiency of cycling was studied. Efficiency was measured in terms of distance (km) pedalled on an ergocycle until exhaustion. The experimenter selected three knee-joint angles of particular interest: 50, 70, and 90 degrees. Ten subjects were randomly allocated to each angel. The drag of the ergocycle was kept constant at 14.7N, and subjects were instructed to pedal at a constant speed of 20 km/h until exhaustion.

1. Carry out an initial data analysis to assess whether there are any aspects of the data that might be a cause for concern in later analyses.
2. Calculate the appropriate analysis of variance table for the data.

3. Use a variety of multiple comparison tests to explore differences between the population means for each angle and compare their results.

50	70	90
8.4	10.6	3.2
7.0	7.5	4.2
3.0	5.1	3.1
8.0	5.6	6.9
7.8	10.2	7.2
3.3	11.0	3.5
4.3	6.8	3.1
3.6	9.4	4.5
8.0	10.4	3.8
6.8	8.8	3.6

Exercise 5.3: Hypertension Data

The data in the SAS data set **hypertension** are from a study described by Maxwell and Delaney (1990) in which the effects of three possible treatments for hypertension were investigated. The details of the treatments are as follows:

Treatment	Description	Levels
drug	Medication	drug X, drug Y, drug Z
biofeed	psychological feedback	present, absent
diet	special diet	present, absent

All 12 combinations of the three treatments were included in a $3 \times 2 \times 2$ design. Seventy-two subjects suffering from hypertension were recruited to the study, with six being randomly allocated to each of 12 treatment combinations. Blood pressure measurements were made on each subject after treatment, leading to the data below.

		Special diet	
Biofeedback	**Drug**	**No**	**Yes**
Present	X	170 175 165 180 160 158	161 173 157 152 181 190
	Y	186 194 201 215 219 209	164 166 159 182 187 174
	Z	180 187 199 170 204 194	162 184 183 156 180 173
		173 194 197 190 176 198	164 190 169 164 176 175
Absent	X	189 194 217 206 199 195	171 173 196 199 180 203
	Y	202 228 190 206 224 204	205 199 170 160 179 179
	Z		

1. Find the analysis of variance table for the data and interpret the results.
2. Construct appropriate graphics as an aid to this interpretation.
3. Reanalyse the data after taking a log transformation and compare the results with those in Step 1.

Exercise 5.4: Genotypes Data

The data in the **genotypes** data set are from a foster feeding experiment with rat mothers and litters of four different genotypes: A, B, I, and J. The measurement is the litter weight (in grams) after a trial feeding period. Investigate the effect of genotype of mother and litter on litter weight.

Litter Genotype	Mother Genotype	Weight
A	A	61.5
A	A	68.2
A	A	64.0
A	A	65.0
A	A	59.7
A	B	55.0
A	B	42.0
A	B	60.2
A	I	52.5
A	I	61.8

Litter Genotype	Mother Genotype	Weight
A	I	49.5
A	I	52.7
A	J	42.0
A	J	54.0
A	J	61.0
A	J	48.2
B	J	39.6
B	A	60.3
B	A	51.7
B	A	49.3
B	A	48.0
B	B	50.8
B	B	64.7
B	B	61.7
B	B	64.0
B	B	62.0
B	I	56.5
B	I	59.0
B	I	47.2
B	I	53.0
B	J	51.3
B	J	40.5
I	A	37.0
I	A	36.3
I	A	68.0
I	B	56.3
I	B	69.8
I	B	67.0
I	I	39.7
I	I	46.0
I	I	61.3
I	I	55.3

Litter Genotype	Mother Genotype	Weight
I	I	55.7
I	J	50.0
I	J	43.8
I	J	54.5
J	A	59.0
J	A	57.4
J	A	54.0
J	A	47.0
J	B	59.5
J	B	52.8
J	B	56.0
J	I	45.2
J	I	57.0
J	I	61.4
J	J	44.8
J	J	51.5
J	J	53.0
J	J	42.0
J	J	54.0

Chapter 6: Multiple Linear Regression

6.1 Introduction

In this chapter, we will discuss how to analyze data in which there are continuous response variables and a number of explanatory variables that can be associated with the response variable. The aim is to build a statistical model that allows us to discover which of the explanatory variables are of most importance in determining the response. The statistical topics covered are

- Multiple regression
- Interpretation of regression coefficients
- Regression diagnostics

6.2 Multiple Linear Regression

The data shown in Table 6.1 were collected in a study to investigate how price and temperature influenced consumption of ice cream. Here the response variable is consumption of ice cream, and we have two explanatory variables, price and temperature. The aim is to fit a suitable statistical model to the data that allows us to determine how consumption of ice cream is affected by the other two variables.

Table 6.1: Ice Cream Consumption Measured Over 30 4-Week Periods

Observation	Consumption	Price	Mean Temperature
1	.386	.270	41
2	.374	.282	56
3	.393	.277	63
4	.425	.280	68
5	.406	.272	69
6	.344	.262	65
7	.327	.275	61
8	.288	.267	47
9	.269	.265	32
10	.256	.277	24
11	.286	.282	28
12	.298	.270	26
13	.329	.272	32
14	.318	.287	40
15	.381	.277	55
16	.381	.287	63
17	.470	.280	72
18	.443	.277	72
19	.386	.277	67
20	.342	.277	60
21	.319	.292	44
22	.307	.287	40
23	.284	.277	32
24	.326	.285	27
25	.309	.282	28

Observation	Consumption	Price	Mean Temperature
26	.359	.265	33
27	.376	.265	41
28	.416	.265	52
29	.437	.268	64
30	.548	.260	71

6.2.1 The Ice Cream Data: An Initial Analysis Using Scatter Plots

The ice cream data in Table 6.1 are in the data set **sasue.icecream**.

Some scatter plots of the three variables will be helpful in an initial examination of the data. For the scatter plots:

1. Open **Tasks ▶ Graph ▶ Scatter Plot**.
2. Under **Data ▶ Data**, add **sasue.icecream**.
3. Under **Data ▶ Roles**, assign **consumption** as the **y variable** and **price** as the **x variable**.
4. Under **Options ▶ X axis**, deselect **Show grid lines** and do the same for the y axis**.**
5. Click **Run**.

Repeat this with **temperature** as the x variable.

The resulting scatter plots are shown in Figures 6.1 and 6.2. The plots suggest that temperature is more influential than price in determining ice cream consumption, with consumption increasing markedly as temperature increases.

Figure 6.1: Scatter Plot of Ice Cream Consumption Against Price

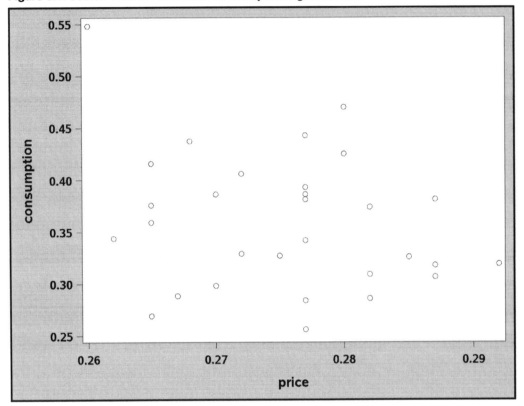

Figure 6.2: Scatter Plot of Ice Cream Consumption Against Temperature

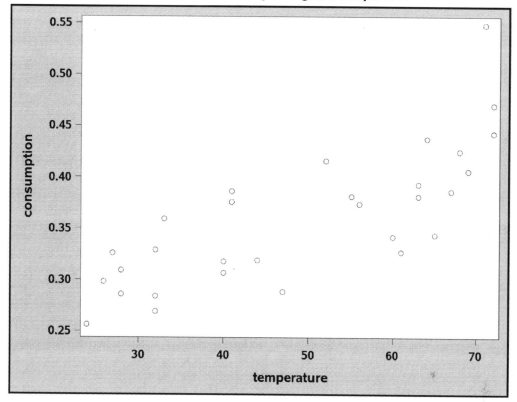

6.2.2 Ice Cream Sales: Are They Most Affected by Price or Temperature? How to Tell Using Multiple Regression

In Chapter 4, we examined the simple linear regression model that allows the effect of a single explanatory variable on a response variable to be assessed. We now need to extend the simple linear regression model to situations where there is more than a single explanatory variable. For continuous response variables, a suitable model is often *multiple linear regression*, which mathematically can be written as:

$$y_i = \beta_0 + \beta_1 x_{i1} + \beta_2 x_{i2} + \ldots\ldots + \beta_q x_{iq} + \varepsilon_i$$

where $y_i, i = 1, n$ are the observed values of the response variable and $x_{i1}, x_{i2}, \ldots, x_{iq}$, $i = 1, n$ are the observed values of the q explanatory variables; n is the number of observations in the sample; and the ε_i terms represent the errors in the model. The *regression coefficients,*

$\beta_1, \beta_2, ..., \beta_q$, give the amount of change in the response variable associated with a unit change in the corresponding explanatory variable, *conditional* on the other explanatory variables remaining unchanged. The regression coefficients are estimated from the sample data by *least squares*. For full details, see, for example, Everitt (1996).

The error terms in the model are each assumed to have a normal distribution with mean zero and variance σ^2. The assumed normal distribution for the error terms in the model implies that, for given values of the explanatory variables, the response variable is normally distributed with a mean that is a linear function of the explanatory variables and a variance that is not dependent on the explanatory variables. The variation in the response variable can be partitioned into a part due to regression on the explanatory variables and a residual. The various terms in the partition of the variation of the response variable can be arranged in an analysis of variance table, and the F-test of the equality of the regression variance (or *mean square*) and the residual variance gives a test of the hypothesis that there is no regression on the explanatory variables (that is, the hypothesis that all the population regression coefficients are 0). Further details are available in Everitt (1996).

We now fit the multiple regression model to the ice cream data--here there are two explanatory variables, temperature and price:

1. Open **Tasks ▶ Statistics ▶ Linear Regression**.
2. Under **Data ▶ Data**, add **sasue.icecream**.
3. Under **Data ▶ Roles**, add **consumption** as the **dependent variable** and **price** and **temperature** as **continuous variables**.
4. Under **Model ▶ Model Effects**, select both **price** and **temperature** (Ctrl-Click on both) and click the **Add** button.
5. Under **Options ▶ Plots ▶ Diagnostic and Residual Plots**, we note that **Diagnostic plots** and **Residuals for each explanatory variable** are selected by default.
6. Click **Run**.

We begin with a discussion of the tabular results, which are shown in Output 6.1. The F-test in the analysis of variance table takes the value 23.27 and has an associated *p*-value that is very small. Clearly the hypothesis that both regression coefficients are 0 is not tenable. The multiple correlation coefficient gives the correlation between the observed values of the response variable (ice cream consumption) and the values predicted by the fitted model; the square of the coefficient (0.63) gives the proportion of the variance in ice cream consumption accounted for by price and temperature. The negative regression coefficient for price indicates that, for a given mean temperature, consumption decreases with increasing price. But as was suggested by the two scatter plots in Subsection 6.3.1, only the estimated regression coefficient associated with temperature is statistically significant. The regression coefficient of consumption on temperature is estimated to be 0.00303, with an estimated standard error of 0.00047. A one-degree rise in temperature is estimated to increase consumption by 0.00303 units, conditional on price.

Output 6.1: Results from Applying the Multiple Regression Model to the Ice Cream Consumption Data

Number of Observations Read	30
Number of Observations Used	30

Analysis of Variance					
Source	DF	Sum of Squares	Mean Square	F Value	Pr > F
Model	2	0.07943	0.03972	23.27	<.0001
Error	27	0.04609	0.00171		
Corrected Total	29	0.12552			

Root MSE	0.04132	R-Square	0.6328
Dependent Mean	0.35943	Adj R-Sq	0.6056
Coeff Var	11.49484		

Parameter Estimates					
Variable	DF	Parameter Estimate	Standard Error	t Value	Pr > |t|
Intercept	1	0.59655	0.25831	2.31	0.0288
price	1	-1.40176	0.92509	-1.52	0.1413
temperature	1	0.00303	0.00046995	6.45	<.0001

6.2.3 Diagnosing the Multiple Regression Model Fitted to the Ice Cream Consumption Data: The Use of Residuals

As with the simple linear regression model described in Chapter 4, an important final stage when fitting a multiple regression model is to investigate whether the assumptions made by the model, such as constant variance and normality of error terms, are reasonable. This involves the consideration of suitable plots of residuals, as discussed in Chapter 4, and perhaps other diagnostic plots as well.

So let us turn now to the plots that were produced by the linear regression task for the ice cream data, which are shown in Figures 6.3 and 6.4. Looking at the latter plot first, we see that there are no apparent patterns in the plots that would indicate problems with the fitted model; the two plots look similar to the idealized residual plot in Figure 4.6 (a) given in Chapter 4.

Figure 6.3 gives a number of *diagnostic plots*. The first plot in the first row of plots is simply a plot of the basic residuals against the predicted value of ice cream consumption. Again, no distinct pattern, for example, like that in Figure 4.6 (b) is apparent. The second plot in the first row is similar to the first but uses what are known as *studentized residuals* rather than the basic residual described earlier. Studentized residuals are described explicitly in Der and Everitt (2013); a form of *standardized residual*, they are needed because the simple residuals can be shown to have unequal variances and to be slightly correlated. This correlation makes them less than ideal for detecting problems with fitted models. Details are given in Rawlings et al. (2001), where an account of *leverage* (see the third plot in the first row) and *Cook's distance* (see the third plot in the second row) can also be found.

For the leverage plot, points should ideally lie in the -2 to 2 band shown on the plot; here there are a couple of points lying outside this band, but on the whole the plot is satisfactory. In Cook's distance index plot, values greater than one give cause for concern; there are no such values here. The first plots in the second and third rows can be used to check the normality assumption in the multiple linear regression model, the first being a *normal probability plot* (see Der and Everitt, 2013) of the residuals and the second a histogram of the residuals showing a fitted normal distribution. Both plots suggest that the normality assumption is justified for these data.

Figure 6.3: Diagnostic and Residual Plots for the Ice Cream Consumption Data

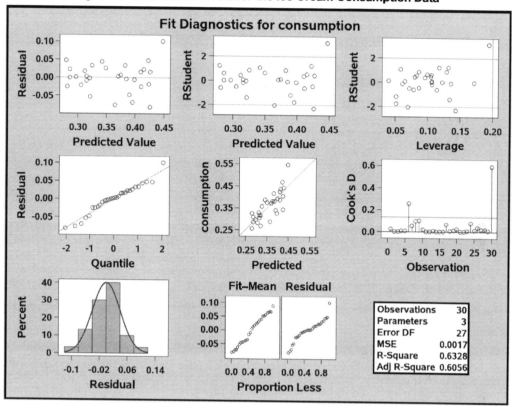

Figure 6.4: Residual Plots for the Ice Cream Consumption Data

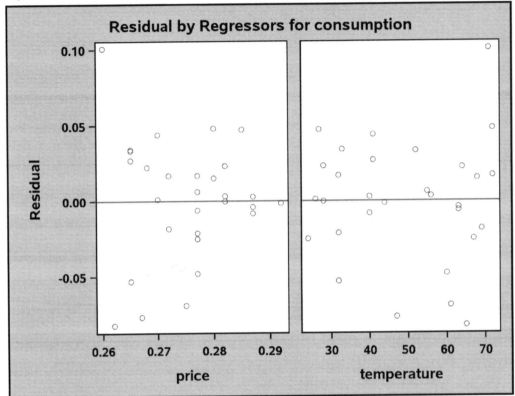

6.2.4 A More Complex Example of the Use of Multiple Linear Regression

Weather modification, or *cloud seeding,* is the treatment of individual clouds or storm systems with various inorganic or organic materials in the hope of achieving an increase in rainfall. Introduction of such material into a cloud that contains super cooled water (that is, liquid water colder than 0 °C) has the aim of inducing freezing, with the consequent ice particles growing at the expense of liquid droplets and becoming heavy enough to fall as rain from the clouds that otherwise would produce none.

The data in Table 6.2 were collected in the summer of 1975 from an experiment to investigate the use of massive amounts of silver iodine (100 to 1,000 grams per cloud) in cloud seeding to increase rainfall (Woodley et al., 1977). In the experiment, which was conducted in an area of Florida, 24 days were judged suitable for seeding on the basis that a measured suitability criterion was met. The suitability criterion (S-NE) that is defined in detail in Woodley et al. biases the decision for experimentation against naturally rainy days. On thus-defined suitable days, a decision was made at random about whether to seed or not. The aim in analysing the cloud seeding data is to see how rainfall is related to the other variables and, in particular, to determine the effectiveness of seeding.

Table 6.2: Cloud Seeding Data

Seeding	Time	S-NE	Cloud Cover	Prewetness	Echo Motion	Rainfall
0	0	1.75	13.40	0.274	1	12.85
1	1	2.70	37.90	1.267	0	5.52
1	3	4.10	3.90	0.198	1	6.29
0	4	2.35	5.30	0.526	0	6.11
1	6	4.25	7.10	0.250	0	2.45
0	9	1.60	6.90	0.018	1	3.61
0	18	1.30	4.60	0.307	0	0.47
0	25	3.35	4.90	0.194	0	4.56
0	27	2.85	12.10	0.751	0	6.35
1	28	2.20	5.20	0.084	0	5.06
1	29	4.40	4.10	0.236	0	2.76
1	32	3.10	2.80	0.214	0	4.05
0	33	3.95	6.80	0.796	0	5.74
1	35	2.90	3.00	0.124	0	4.84
1	38	2.05	7.00	0.144	0	11.86
0	39	4.00	11.30	0.398	0	4.45
0	53	3.35	4.20	0.237	1	3.66
1	55	3.70	3.30	0.960	0	4.22
0	56	3.80	2.20	0.230	0	1.16
1	59	3.40	6.50	0.142	1	5.45
1	65	3.15	3.10	0.073	0	2.02
0	68	3.15	2.60	0.136	0	0.82
1	82	4.01	8.30	0.123	0	1.09
0	83	4.65	7.40	0.168	0	0.28

6.2.5 The Cloud Seeding Data: Initial Examination of the Data Using Box Plots and Scatter Plots

For the cloud seeding data, we will construct box plots of the rainfall in each category of the dichotomous explanatory variables (**seeding** and **echomotion**) and scatter plots of rainfall against each of the continuous explanatory variables (**cloudcover**, **sne**, **prewetness**, and **time**).

For the box plots:

1. Open **Tasks ▶ Graph ▶ Box Plot**.
2. Under **Data ▶ Data**, select **sasue.cloud**.
3. Under **Data ▶ Roles**, add **rainfall** as the **analysis variable** and **seeding** as the **category** variable.
4. Click **Run**.

Repeat with **echomotion** as the category variable.

For the scatter plots:

1. Open **Tasks ▶ Graph ▶ Scatter Plot**.
2. Under **Data ▶ Data**, select **sasue.cloud**.
3. Under **Data ▶ Roles**, add **rainfall** as the **y variable** and **time** as the **x variable**.
4. Click **Run**.

Repeat with **sne**, **cloudcover**, and **prewetness** as x variables.

The plots are shown in Figures 6.5 to 6.10. Both the box plots and the scatter plots show some evidence of two outliers. In particular, the scatter plot of rainfall against cloud cover suggests one very clear outlying observation, which, on inspection, turns out to be the second observation in the data set. For the time being, we shall not remove any observations, but simply bear in mind during the modelling process to be described later that outliers can cause difficulties.

Figure 6.5: Box Plots of Rainfall for Seeding and Not Seeding Days

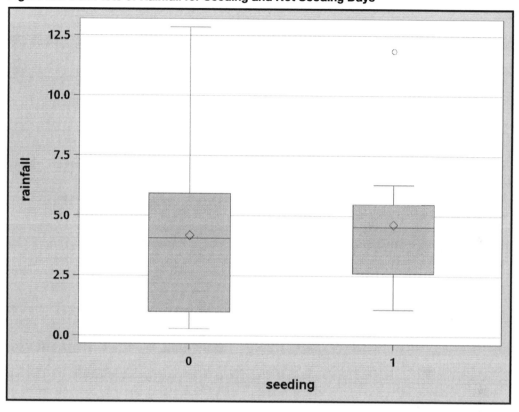

Figure 6.6: Box Plots of Rainfall for Moving and Stationary Echo Motion

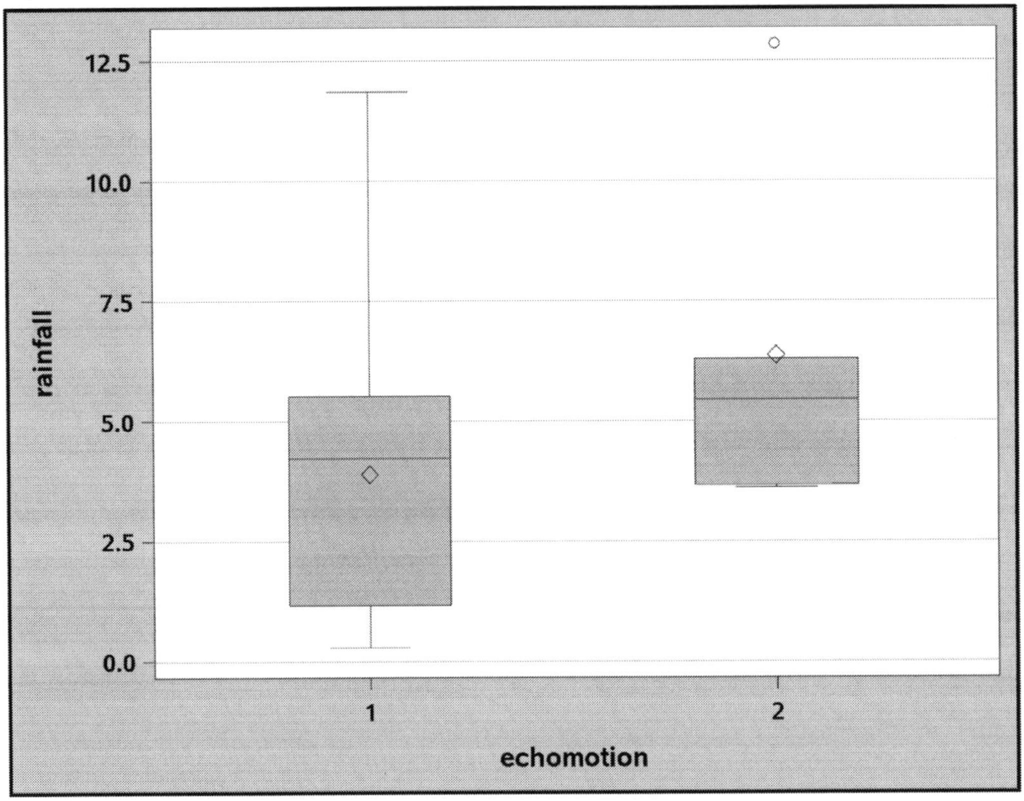

Figure 6.7: Scatter Plot of Rainfall Against Time

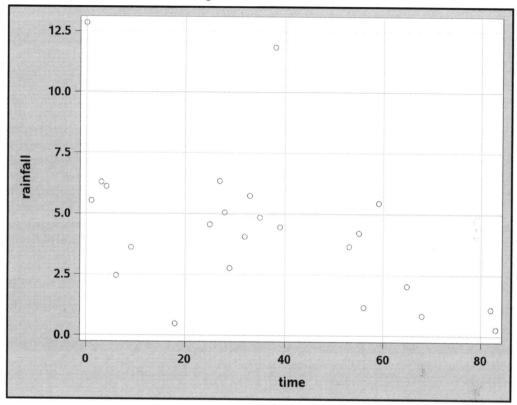

Figure 6.8: Scatter Plot of Rainfall Against S-NE Criterion

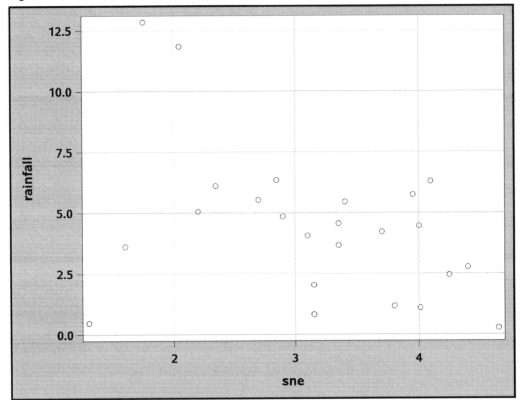

Figure 6.9: Scatter Plot of Rainfall Against Cloud Cover

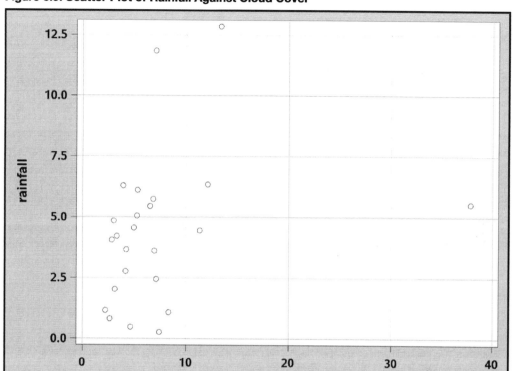

Figure 6.10: Scatter Plot of Rainfall Against Prewetness

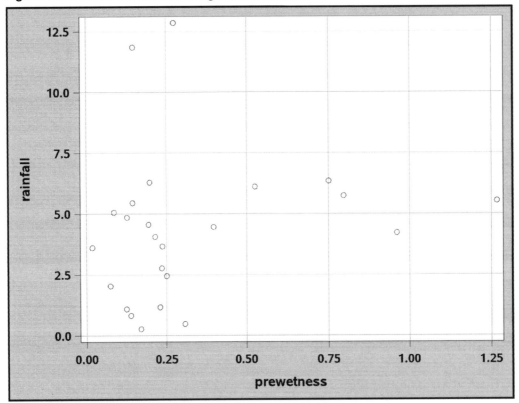

6.2.6 When Is Cloud Seeding Best Carried Out? How to Tell Using Multiple Regression Models Containing Interaction Terms

For the cloud seeding data, one thing to note about the explanatory variables is that two of them, **seeding** and **echomotion**, are binary variables. Should such an explanatory variable be allowed in a multiple regression model? In fact, there is really no problem in including such variables in the model because, in a strict sense, the explanatory variables are assumed to be fixed rather than random. (In practice, of course, fixed explanatory variables are rarely the case, so the results from a multiple regression analysis are interpreted as being conditional on the observed values of the explanatory variables; this rather arcane point is discussed in more detail in Everitt, 1996. It is only the response variable that is considered to be random.)

In the cloud seeding example, there are theoretical reasons to consider a particular model for the data (see Woodley et al., 1977)—namely, one in which the effect of some of the other explanatory variables is modified by seeding. So the model that we will consider is one that allows *interaction terms* for seeding with each of the other explanatory variables except **time**:

1. Open **Tasks ▶ Statistics ▶ Linear Regression**.
2. Under **Data ▶ Data**, select **sasue.cloud**.
3. **Under Data ▶ Roles**, add **rainfall** as the **dependent variable** and treat the remaining variables as **continuous variables**.
4. Under **Model**, select all the variables and click **Add** to enter their main effects.

To enter the required interactions with **seeding**, first click on **seeding**, then Ctrl-Click on the other variable in the interaction (for example, **time**), and then click the **Cross** button. Repeat this for each interaction with **seeding**. When all the terms have been entered, the model pane should look like Display 6.1.

Display 6.1: Model Pane for Cloud Seeding Data

The results of fitting the specified model are shown in Output 6.2. To begin, let us look at the least squares summary table part of the output. This shows how the fit of the regression model changes as the variables are entered into the model in the order specified. The measure of fit used is what is known as the *Schwarz Bayesian Criterion* (*SBC*); this measure is defined explicitly in Everitt and Skrondal (2010), but, in simple terms, the measure takes into account both the statistical goodness of fit of a model and the number of parameters that have to be estimated to achieve this particular degree of fit by imposing a penalty for increasing the number of parameters.

Essentially, the SBC tries to find a model with the fewest number of parameters that still provides an adequate fit to the data. Lower values of the SBC indicate the preferred model. In Output 6.2, this criterion indicates that the model that includes **seeding** x **cloudcover** interaction and all the variables entered before this term is the preferred model. But it has to be remembered that just as the explanatory variables are not independent of one another, neither are the estimated regression coefficients (see the next section for more discussion of this point). Changing the order in which variables are considered in the least squares summary table might lead to a slightly different preferred model being chosen by the SBC.

Moving on through the output, we find the analysis of variable table in which the F-test assesses the hypothesis that all the regression coefficients in the model are 0 (that is, none of the explanatory variables affect the response variable, **rainfall**). The significance value associated with the test is 0.024, implying that this very general hypothesis can be rejected. Underneath the analysis of variance table, we find (amongst other things that we will ignore for the moment) the value of the square of the multiple correlation coefficient, which gives the proportion of the variance of the response variable accounted for by the explanatory variables. Here the value is 0.72.

Also, in this part of the output the value for the adjusted R-square appears; this is a modification of R-square that adjusts for the number of explanatory variables in a model. The adjusted R-square does not have the same interpretation as R-square and is essentially an index of the fit of the model in the same manner as the SBC. We shall see this in the next section, where we discuss how to select a parsimonious regression model for a data set.

Finally, Output 6.2 contains the estimated regression coefficients, their estimated standard errors, and the associated *t*-tests for assessing the hypothesis that the corresponding population regression coefficient is 0.

Output 6.2: Results of Fitting a Multiple Regression Model with Interactions to the Cloud Seeding Data

Data Set	SASUE.CLOUD
Dependent Variable	rainfall
Selection Method	None

Number of Observations Read	24
Number of Observations Used	24

Dimensions	
Number of Effects	11
Number of Parameters	11

Least Squares Summary			
Step	Effect Entered	Number Effects In	SBC
0	Intercept	1	56.6052
1	seeding	2	59.6443
2	time	3	55.9111
3	sne	4	57.3639
4	cloudcover	5	60.2283
5	prewetness	6	63.3540
6	echomotion	7	64.0090
7	seeding*sne	8	60.5224
8	seeding*cloudcover	9	52.4302*
9	seeding*prewetness	10	55.0976
10	seeding*echomotion	11	58.1924
* Optimal Value of Criterion			

Analysis of Variance					
Source	DF	Sum of Squares	Mean Square	F Value	Pr > F
Model	10	159.14600	15.91460	3.27	0.0243
Error	13	63.18889	4.86068		
Corrected Total	23	222.33490			

Root MSE	2.20470
Dependent Mean	4.40292
R-Square	0.7158
Adj R-Sq	0.4972
AIC	71.23379
AICC	99.59743
SBC	58.19239

Parameter Estimates					
Parameter	DF	Estimate	Standard Error	t Value	Pr > |t|
Intercept	1	-3.499055	4.063201	-0.86	0.4048
seeding	1	16.245153	5.521635	2.94	0.0114
time	1	-0.044974	0.025053	-1.80	0.0959
sne	1	0.419814	0.844530	0.50	0.6274
cloudcover	1	0.387862	0.217855	1.78	0.0984
prewetness	1	4.108342	3.601007	1.14	0.2745
echomotion	1	3.152814	1.932526	1.63	0.1268
seeding*sne	1	-3.197190	1.267072	-2.52	0.0254
seeding*cloudcover	1	-0.486255	0.241060	-2.02	0.0648
seeding*prewetness	1	-2.557067	4.480896	-0.57	0.5780
seeding*echomotion	1	-0.562218	2.644300	-0.21	0.8349

The tests of the interactions in the model suggest that the interaction of seeding with the S-NE criterion significantly affects rainfall. A suitable graph will help in the interpretation of the significant seeding x S-NE criterion interaction:

1. Open **Tasks ▶ Graph ▶ Scatter Plot**.
2. Under **Data ▶ Data**, add **sasue.cloud**.
3. Under **Data ▶ Roles**, add **rainfall** as the **y variable**, **sne** as the **x variable**, and **seeding** as the **group variable**.
4. Under **Data ▶ Fit Plots**, select **Regression**.
5. Click **Run**.

The result is shown in Figure 6.11.

Figure 6.11: Scatter Plot of Rainfall versus S-NE for Seeding and Non-Seeding Days

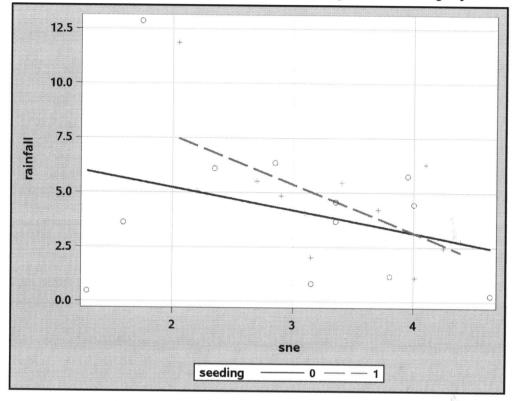

The plot suggests that for smaller S-NE values, seeding produces greater rainfall than not seeding, whereas for larger S-NE values, it tends to produce less. The crossover point is at an S-NE value of approximately 4, which might suggest that, for the most success, seeding should be applied when the S-NE criterion is less than 4.

6.3 Identifying a Parsimonious Regression Model

A multiple regression model begins with a set of observations on a response variable and a number of explanatory variables. After an initial analysis has established that some, at least, of the explanatory variables are predictive of the response, the question arises whether a subset of the explanatory variables might provide a simpler model that is essentially as useful as the full model in predicting, or explaining, the response. It might be thought that such a model could easily be chosen by simply examining the tests of the hypotheses that the regression coefficients associated with each of the explanatory variables in the full model are 0 and then selecting only those variables whose regression coefficients are significantly different from 0 at some particular significance level. This is, indeed, simple, but sadly it is not a procedure that can be recommended. The problem is that the explanatory variables are rarely independent of each other and, consequently, neither are the estimated regression coefficients; dropping a particular explanatory variable from a model, for example, and then fitting the reduced model will likely lead to different estimated regression coefficients for the variables in both models.

For this reason, a number of methods for choosing the best set of explanatory variables have been developed; *forward selection*, for example, uses some numerical criterion to decide whether a variable should be added to an existing model and *backward elimination* uses such a criterion to decide whether a variable in an existing model can be dropped from that model. And *stepwise regression* uses a combination of these two procedures. For a detailed account of variable selection methods, see Der and Everitt (2013). Here we shall simply illustrate the use of backward elimination using the data shown in Table 6.3, which arise from asking healthy active females to run on a treadmill until unable to go any further. The explanatory variables recorded are the duration of the run in seconds, the maximum heart rate during the exercise (bpm), the participant's age in years, her height (cm), and her weight (kg). Also measured is the participant's volume of oxygen used per minute per kilogram of body weight, VO2MAX. Full details of the data are given in Bruce et al. (1973) and in van Belle et al. (2004). Interest lies in finding a model for predicting VO2MAX from the five explanatory variables.

Table 6.3: Exercise Data for Healthy Active Females

Duration	VO2MAX	Heart Rate	Age	Height	Weight
660	38.1	184	23	177	83
628	38.4	183	21	163	52
637	41.7	200	21	174	61
575	33.5	170	42	160	50
590	28.6	188	34	170	68
600	23.9	190	43	171	68
562	29.6	190	30	172	63
495	27.3	180	49	157	53
540	33.2	184	30	178	63
470	26.6	162	57	161	63
408	23.6	188	58	159	54
387	23.1	170	51	162	55
564	36.6	184	32	165	57
603	35.8	175	42	170	53
420	28.0	180	51	158	47
573	33.8	200	46	161	60
602	33.6	190	37	173	56
430	21.0	170	50	161	62

Duration	VO2MAX	Heart Rate	Age	Height	Weight
508	31.2	158	65	165	58
565	31.2	186	40	154	69
464	23.7	166	52	166	67
495	24.5	170	40	160	58
461	30.5	188	52	162	64
540	25.9	190	47	161	72
588	32.7	194	43	164	56
498	26.9	190	48	176	82
483	24.6	190	43	165	61
554	28.8	188	45	166	62
521	25.9	184	52	167	62
436	24.4	170	52	168	62
398	26.3	168	56	162	66
366	23.2	175	56	159	56
439	24.6	156	51	161	61
549	28.8	184	44	154	56
360	19.6	180	56	167	79
566	31.4	184	40	165	56
407	26.6	156	53	157	52
602	30.6	194	52	161	65
488	27.5	190	40	178	64
526	30.9	188	55	162	61
524	33.9	164	39	166	59
562	32.3	185	57	168	68
496	26.9	178	46	156	53

We will fit multiple regression models to these data, applying backward selection to find a parsimonious model for the data; the significance level of the *t*-test for assessing whether a population regression coefficient is 0 has the selection criterion for removing variables from the starting model, which includes all five explanatory variables. **The required SAS code is as follows**:

1. Open **Tasks ▶ Statistics ▶ Linear Regression**.
2. Under **Data ▶ Data**, select **sasue.treadmill**.

3. Under **Data ▶ Roles**, add **vo2max** as the dependent variable and treat the remaining variables as continuous variables.

4. Under **Model ▶ Model Effects**, select all the variables and click **Add** to enter their main effects.

5. Under **Selection ▶ Model Selection**, choose **Backward elimination** as the **Selection method** and **Significance level** as the **Criterion to add or remove effects**.

6. Click **Run**.

The result is shown in Output 6.3. The first part of this output indicates the selection criterion being used, which we have chosen to be the significance level as explained above and the value of the threshold significance level below which variables will not be removed for an existing model. The next part of the output shows that the variables for heart rate, height, and age can be removed from the full model, leaving the chosen model with just the two explanatory variables, duration and weight. The R-square value for the selected model shows that duration and weight account for about 65% of the variation in VO2MAX.

The fit criteria plots show how various measures of fit change as variables are removed from the full model. We have already met the SBC and the adjusted R-square measures; the AIC (Akaike's information criterion) and AIIC (a slightly amended version of the Akaike information criterion) are similar to SBC, but have different weightings for fit against the number of parameters. They are defined explicitly in Everitt and Skrondal (2010) and Hurvich and Tsai (1989). Examining these plots, we see that SBC selects the same model as the backward selection approach using the significance level criterion, but the other three fit indices chose a model in which age as well as duration and weight are included. Here we shall keep with the simpler model and move on to look at the other graphical output, which include the diagnostic plots (Figures 6.12 through 6.15) previously discussed when dealing with the ice cream consumption data. The only concerning finding in these plots is perhaps that there are a small number of outliers, which may distort the fitted models in some way. We leave it as an exercise for readers to identify these outliers and then repeat the model-fitting procedure described in this section with the identified outliers removed.

Output 6.3: Results from Applying Backward Selection to the Data in Table 6.3

Data Set	SASUE.TREADMILL
Dependent Variable	VO2MAX
Selection Method	Backward
Select Criterion	Significance Level
Stop Criterion	Significance Level
Stay Significance Level (SLS)	0.05
Effect Hierarchy Enforced	None

Number of Observations Read	43
Number of Observations Used	43

Dimensions	
Number of Effects	6
Number of Parameters	6

Backward Selection Summary				
Step	Effect Removed	Number Effects In	F Value	Pr > F
0		6		
1	Heart_rate	5	0.47	0.4960
2	Height	4	0.69	0.4114
3	Age	3	2.54	0.1187

Selection stopped because the next candidate for removal has SLS < 0.05.

Stop Details				
Candidate For	Effect	Candidate Significance	Compare Significance	
Removal	Weight	0.0360 <	0.0500	(SLS)

Figure 6.12: Plot of Fit Criteria Against Model Chosen for the Exercise Data

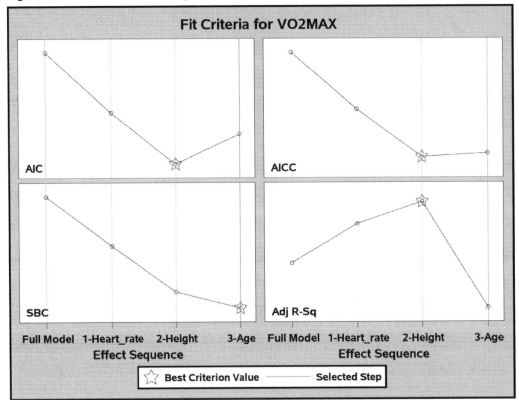

The selected model is the model at the last step (Step 3).

Effects: Intercept Duration Weight

Analysis of Variance				
Source	DF	Sum of Squares	Mean Square	F Value
Model	2	676.93480	338.46740	38.49
Error	40	351.73217	8.79330	
Corrected Total	42	1028.66698		

Root MSE	2.96535
Dependent Mean	29.05349
R-Square	0.6581
Adj R-Sq	0.6410
AIC	141.37181
AICC	142.42444
SBC	101.65541

Parameter Estimates				
Parameter	DF	Estimate	Standard Error	t Value
Intercept	1	10.300264	4.506680	2.29
Duration	1	0.051500	0.005942	8.67
Weight	1	-0.126589	0.058329	-2.17

Figure 6.13: Plot of Observed and Fitted Value for the Exercise Data

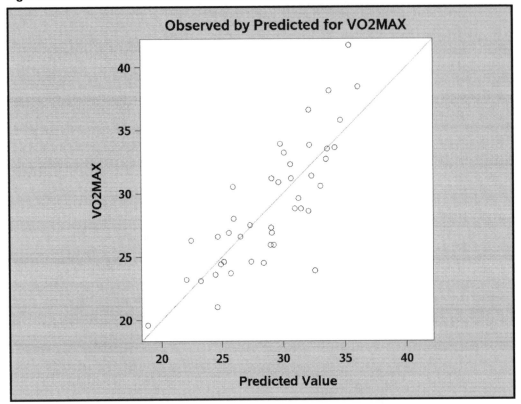

Figure 6.14: Diagnostic and Residual Plots for the Exercise Data

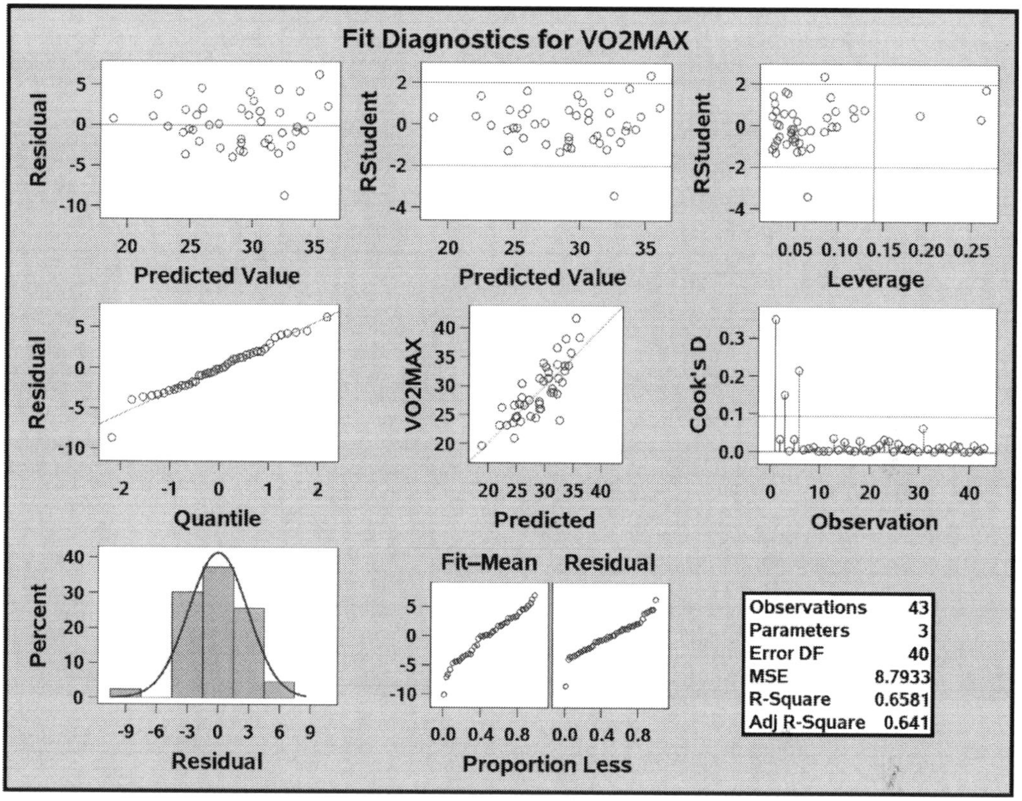

Figure 6.15: Residual Plots for the Exercise Data

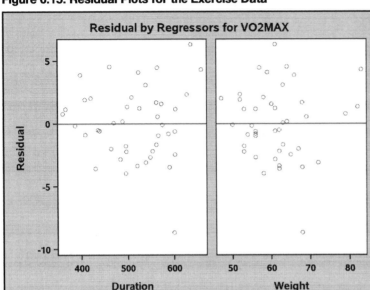

Automatic model selection methods must be used with care, and the researcher using them should approach the final model selected with an appropriate degree of scepticism. Agresti (1996) nicely summarizes the potential problems:

> *Computerized variable selection procedures should be used with caution. When one considers a large number of terms for potential inclusion in a model, one or two of them that are not really important might look impressive simply due to chance. For example when all the true effects are weak, the largest sample effect might substantially overestimate its true effect. In addition, it makes sense to include variables of special interest in a model and report their estimated effects even if they are not statistically significant at some level.*

(See McKay and Campbell, 1982a, 1982b for some more thoughts on automatic selection methods in regression.)

6.4 Exercises

Exercise 6.1: Ice Cream Data

For the ice cream consumption data, investigate what happens when you fit a simple linear regression of consumption on **price** and then add **temperature** to the model. Repeat this exercise fitting first **temperature** and then adding **price**.

Exercise 6.2: Cloud Data

Repeat the analysis of the cloud seeding data after removing observations 1 and 15. Compare your results with those given in the text.

Exercise 6.3: Fat Data

The data in the **fat** data set are taken from a study investigating a new method of measuring body composition and give the body fat percentage, age, and sex for 20 normal adults aged between 23 and 61 years.

1. Construct a scatter plot of the percentage of fat against the age, labeling the points according to the sex.
2. Fit a multiple regression model to the data using **fat** as the response variable and **age** and **sex** as explanatory variables. Interpret the results with the help of a scatter plot showing the essential features of the fitted model.
3. Fit a further model that allows an interaction between **age** and **sex** and again construct a diagram that will help you interpret the results.

Age	Sex	% Fat
23	M	9.5
23	F	27.9
27	M	7.8
27	M	17.8
39	F	31.4
41	F	25.9
45	M	27.4
49	F	25.2
50	F	31.1
53	F	34.7
53	F	42.0
54	F	29.1
56	F	32.5
57	F	30.3
58	F	33.0
58	F	33.8
60	F	41.1
61	F	34.5

Exercise 6.4: Air Pollution Data

The data in the dataset relate to air pollution in 41 US cities. Seven variables are recorded for each of the cities:

1. SO_2 content of air in micrograms per cubic meter
2. Average annual temperature in °F
3. Number of manufacturing enterprises employing 20 or more workers
4. Population size (1970 census) in thousands
5. Average annual wind speed in miles per hour
6. Average annual precipitation in inches
7. Average number of days with precipitation per year

SO2	Temperature	Factories	Population	Wind	Rain	Rainy Days
10	70.3	213	582	6	7.05	36
13	61	91	132	8.2	48.52	100
12	56.7	453	716	8.7	20.66	67
17	51.9	454	515	9	12.95	86
56	49.1	412	158	9	43.37	127
36	54	80	80	9	40.25	114
29	57.3	434	757	9.3	38.89	111
14	68.4	136	529	8.8	54.47	116
10	75.5	207	335	9	59.8	128
24	61.5	368	497	9.1	48.34	115
110	50.6	3344	3369	10.4	34.44	122
28	52.3	361	746	9.7	38.74	121
17	49	104	201	11.2	30.85	103
8	56.6	125	277	12.7	30.58	82
30	55.6	291	593	8.3	43.11	123
9	68.3	204	361	8.4	56.77	113
47	55	625	905	9.6	41.31	111
35	49.9	1064	1513	10.1	30.96	129
29	43.5	699	744	10.6	25.94	137
14	54.5	381	507	10	37	99

SO2	Temperature	Factories	Population	Wind	Rain	Rainy Days
56	55.9	775	622	9.5	35.89	105
14	51.5	181	347	10.9	30.18	98
11	56.8	46	244	8.9	7.77	58
46	47.6	44	116	8.8	33.36	135
11	47.1	391	463	12.4	36.11	166
23	54	462	453	7.1	39.04	132
65	49.7	1007	751	10.9	34.99	155
26	51.5	266	540	8.6	37.01	134
69	54.6	1692	1950	9.6	39.93	115
61	50.4	347	520	9.4	36.22	147
94	50	343	179	10.6	42.75	125
10	61.6	337	624	9.2	49.1	105
18	59.4	275	448	7.9	46	119
9	66.2	641	844	10.9	35.94	78
10	68.9	721	1233	10.8	48.19	103
28	51	137	176	8.7	15.17	89
31	59.3	96	308	10.6	44.68	116
26	57.8	197	299	7.6	42.59	115
29	51.1	379	531	9.4	38.79	164
31	55.2	35	71	6.5	40.75	148
16	45.7	569	717	11.8	29.07	123

Use multiple regression to investigate which of the other variables most determine pollution as indicated by the SO_2 content of the air and find a parsimonious regression model for the data. (Preliminary investigation of the data might be necessary to identify possible outliers and pairs of explanatory variables that are so highly correlated that they can cause problems for model fitting.)

Chapter 7: Logistic Regression

7.1 Introduction

In this chapter, we describe how to deal with data when there is a binary response variable and a number of explanatory variables and you want to see how the explanatory variables affect the response variable. The statistical topics to be covered are

- Regression model for a binary response variable--Logistic regression
- What the logistic regression model tells us--Interpretation of regression coefficients and odds ratios

7.2 Logistic Regression

In a study reported by Pine et al. (1983), patients with intra-abdominal sepsis severe enough to require surgery were followed to determine the incidence of organ failure or death (from sepsis). A number of explanatory variables were also recorded for each patient and the aim was to determine how these explanatory variables affected survival. The data are also given in van Belle et al. (2004). Here we shall use the data from 50 of the original 106 patients in the study; these data are shown in Table 7.1.

Table 7.1: Survival Status of Patients Following Surgery

ID	Survival Status	Shock	Malnutrition	Alcoholism	Age	Bowel Infarction
1	0	0	0	0	56	0
2	0	0	0	0	80	0
3	0	0	0	0	61	0
4	0	0	0	0	26	0
5	0	0	0	0	53	0
6	1	0	1	0	87	0
7	0	0	0	0	21	0
8	1	0	0	1	69	0
9	0	0	0	0	57	0
10	0	0	1	0	76	0
11	1	0	0	1	66	1
12	0	0	0	0	48	0
13	0	0	0	0	18	0
14	0	0	0	0	46	0
15	0	0	1	0	22	0
16	0	0	1	0	33	0
17	0	0	0	0	38	0
18	0	0	0	0	27	0
19	1	1	1	0	60	1
20	0	0	0	0	59	1
21	1	1	0	0	63	1
22	1	1	1	0	70	0
23	1	0	0	1	49	0
24	0	1	0	0	50	0
25	1	1	1	0	70	0
26	1	0	0	1	49	0
27	1	1	0	0	78	1

ID	Survival Status	Shock	Malnutrition	Alcoholism	Age	Bowel Infarction
28	0	0	1	1	60	0
29	0	0	0	1	60	0
30	1	1	0	0	78	1
31	0	0	0	0	28	1
32	0	0	0	0	80	0
33	0	0	0	0	59	1
34	1	0	1	0	50	1
35	1	0	1	0	68	0
36	0	0	0	0	74	1
37	0	0	1	0	27	0
38	1	0	1	1	66	1
39	1	1	1	0	76	0
40	1	0	0	1	70	1
41	0	0	0	0	36	0
42	0	0	0	0	52	1
43	0	0	0	0	30	1
44	1	1	0	1	60	0
45	0	0	1	1	54	0
46	0	0	0	0	65	0
47	1	0	0	0	47	0
48	0	0	1	1	42	0
49	0	0	0	0	22	0
50	0	1	1	0	44	0

Note that shock, malnutrition, alcoholism, and bowel infarction are binary variables coded as 1 if the symptom was present and 0 if it was absent. Survival status is coded as 1 for death and 0 for alive. Age is given in years.

7.2.1 Intra-Abdominal Sepsis: Using Logistic Regression to Answer the Question of What Predicts Survival after Surgery

The multiple regression model considered in the previous chapter is suitable for investigating how a continuous response variable depends on a set of explanatory variables. But can it be adapted to model a *binary response variable*? For example, in Table 7.1, a patient's survival after surgery is such a binary response variable, and we would like to investigate how it is affected by the other variables in the data set. A possible way to proceed is to consider modelling the probability that the binary response takes the value 1 (that is, in our particular example, the probability that a patient dies after surgery for intra-abdominal sepsis). A little thought shows that the multiple regression model cannot help us here. Firstly, the assumption that the response is normally distributed conditional on the explanatory variables is clearly no longer justified. And there is another fundamental problem: the application of the multiple regression model to the probability that the binary response takes the value 1 could lead to fitted values *outside* the range of 0 to 1, clearly unacceptable for the probability being modelled.

7.2.2 Odds

So, with a binary response variable, we need to consider an alternative approach to multiple regression, and the most common alternative is known as *logistic regression*. Here the logarithm of the *odds* of the response variable being 1 (often known as the *logit* transformation of the probability) is modelled as a linear function of the explanatory variables. Odds were discussed in Chapter 3, where we saw that they are simply the ratio of the probability that the binary variable takes the value 1 to the probability that the variable takes the value 0. Representing the probability of a 1 as p, so that the probability of a 0 is $(1-p)$, then the odds are simply given by $p/(1-p)$. So, for example, when tossing an unbiased die, the odds of getting a 6 are 1/6 divided by 5/6, giving the value 1/5. An experienced gambler would say that the odds of a 6 are 5 to 1 against.

But back to the logistic regression model, which in mathematical terms can be written as:

$$\log(\frac{p}{1-p}) = \beta_0 + \beta_1 x_1 + \beta_2 x_2 + \ldots \beta_q x_q$$

where $x_1, x_2, \ldots, x_q$ are the q explanatory variables. Now as p varies between 0 and 1, the logit transformation of p varies between minus and plus infinity, thus removing directly one of the problems mentioned above (that is, an estimated probability outside the range [0, 1]). The logistic model can be rewritten in terms of the probability p as:

$$p = \frac{\exp(\beta_0 + \beta_1 x_1 + \beta_2 x_2 + \ldots \beta_q x_q)}{1 + \exp(\beta_0 + \beta_1 x_1 + \beta_2 x_2 + \ldots \beta_q x_q)}$$

Full details of the distributional assumptions of the model and how the parameters in the model are estimated are given in Der and Everitt (2013), but essentially a binomial distribution is assumed for the response and then the parameters in the model are estimated by maximum likelihood. Below, however, we shall concentrate on how to obtain estimates of the parameters, $\beta_0, \beta_1, \beta_2, \ldots, \beta_q$, using SAS University Edition and how to interpret the estimates after we find them.

7.2.3 Applying the Logistic Regression Model with a Single Explanatory Variable

To begin, we shall apply the logistic regression model to the data in Table 7.1 using **shock** as the single explanatory variable; considering this very simple model should help clarify various aspects of the model before we proceed to consider a more complex model involving all five explanatory variables. But before fitting the logistic model, it will be helpful to look at the cross-classification of survival and shock. This we can find using the following instructions:

1. Open **Tasks ▶ Statistics ▶ Table Analysis**.
2. Under **Data ▶ Data**, add **sasue.sepsis**.
3. Under **Data ▶ Roles**, add **survival** to the **row variables** and **shock** to the **column variables**.
4. Click **Run**.

The edited output just showing the table of interest is given in Output 7.1.

Output 7.1: 2 X 2 Table for Shock and Survival

Table of Survival by Shock			
Survival	**Shock**		
Frequency	**0**	**1**	**Total**
0	30	2	32
1	10	8	18
Total	40	10	50

From this table, we can estimate the probability of death for a patient who has shock as 8/10=0.80 and the corresponding probability for a patient who does not have shock as 10/40=0.25.

Now we can move on and fit the logistic regression model with **shock** as the only explanatory variable. This model can be written explicitly as:

$$\log\left[\frac{p}{1-p}\right] = \beta_0 + \beta_1 \text{Shock}$$

The model can be fitted using the following instructions:

1. Open **Tasks ▶ Statistics ▶ Binary Logistic Regression**.
2. Under **Data ▶ Data**, add **sasue.sepsis**.
3. Under **Data ▶ Roles ▶ Response**, add **survival** and select the **Event of interest** to be 1.
4. Under **Data ▶ Roles ▶ Explanatory Variable ▶ Continuous Variables**, add **shock**.

5. Under **Model** ▶ **Model Effects**, select **shock** and click **Add**.
6. Click **Run**.

Output 7.2: Results from Fitting the Logistic Regression Model to the Survival from Surgery Data with a Single Explanatory Variable, Shock

Analysis of Maximum Likelihood Estimates					
Parameter	DF	Estimate	Standard Error	Wald Chi-Square	Pr > ChiSq
Intercept	1	-1.0986	0.3651	9.0521	0.0026
Shock	1	2.4849	0.8708	8.1425	0.0043

Odds Ratio Estimates			
Effect	Point Estimate	95% Wald Confidence Limits	
Shock	12.000	2.177	66.134

The output of most interest is shown in Output 7.2. Here we find the estimates of the two regression coefficients in the model as $\hat{\beta}_0 = -1.10$ and $\hat{\beta}_1 = 2.48$ (the 'hats' denote sample estimates of the parameters in the model rather than population values). The fitted model is therefore

$$\log\left[\frac{\hat{p}}{1-\hat{p}}\right] = -1.10 + 2.48\text{Shock}$$

So for Shock=0 (no shock), this gives

$$\log\left[\frac{\hat{p}}{1-\hat{p}}\right] = -1.10 \text{ and } \hat{p} = \frac{\exp(-1.10)}{1+\exp(-1.10)} = 0.25$$

And the corresponding calculation for Shock=1 gives $\hat{p} = 0.80$. These two estimates are, of course, equal to those derived previously based on the 2 x 2 contingency table of survival against shock.

Now let us look at another logistic regression model again with a single explanatory variable, the age of the patient; we can do this by making changes to the settings of the previous task:

1. Click the **Binary Logistic Regression** tab.
2. Under **Data ▶ Roles ▶ Explanatory Variables ▶ Continuous Variables**, add **age** and remove **shock**.
3. Under **Model ▶ Model Effects**, select **age** and click **Add**.
4. Click **Run**.

The edited output is shown in Output 7.3. The first part of the output simply gives information about the data set being analysed and the technique used for getting the estimate of the regression coefficients in the model. The statement that the probability modelled is Survival=1 means that it is the probability of death which is modelled, as the variable **survival** takes the value 1 for a patient who dies. Then the estimated regression coefficients are given with their estimated standard errors. Here, of course, it is the regression coefficient for age that is of interest and three tests of the hypothesis that the population regression coefficient takes the value 0 are given. As sample size increases, these different tests lead to the same conclusion--namely, that there is strong evidence that the regression coefficient is not 0. (For a discussion of the details of the three tests, see Collett, 2003.) To help in the interpretation of the fitted model, a confidence interval for the odds ratio is given (see Chapter 3); here, the 95% confidence interval is [1.03, 1.13]. This implies that an extra year of age increases the odds of not surviving surgery for intra-abdominal sepsis between about 3% and 13% over the odds of surviving surgery in this population of patients.

It is helpful to look at a graphic representation of the fitted model, in this case a plot of the estimated probability of dying from surgery against age. We can obtain this plot using the following instructions:

1. Click the **Binary Logistic Regression** tab.
2. Under **Options ▶ Plots ▶ Select Plots to Display**, select **Default & additional plots**.
3. From the resulting list, select **Effect plot**.
4. Click **Run**.

The resulting plot appears in Figure 7.1. This plot clearly demonstrates the effect of aging on the probability of dying from surgery.

Output 7.3: Edited Output from Fitting a Logistic Model to the Data in Table 7.1 with Age as the Single Explanatory Variable

Model Information	
Data Set	SASUE.SEPSIS
Response Variable	Survival
Number of Response Levels	2

Model Information	
Model	binary logit
Optimization Technique	Fisher's scoring

Number of Observations Read	50
Number of Observations Used	50

Probability modeled is Survival=1.

Testing Global Null Hypothesis: BETA=0			
Test	Chi-Square	DF	Pr > ChiSq
Likelihood Ratio	13.7984	1	0.0002
Score	11.8657	1	0.0006
Wald	9.1207	1	0.0025

Analysis of Maximum Likelihood Estimates					
Parameter	DF	Estimate	Standard Error	Wald Chi-Square	Pr > ChiSq
Intercept	1	-4.9362	1.5529	10.1038	0.0015
Age	1	0.0764	0.0253	9.1207	0.0025

Odds Ratio Estimates			
Effect	Point Estimate	95% Wald Confidence Limits	
Age	1.079	1.027	1.134

Figure 7.1: Plot of Predicted Probabilities from Logistic Regression Model with Age as the Single Explanatory Variable Fitted to Data in Table 7.1 Against Age

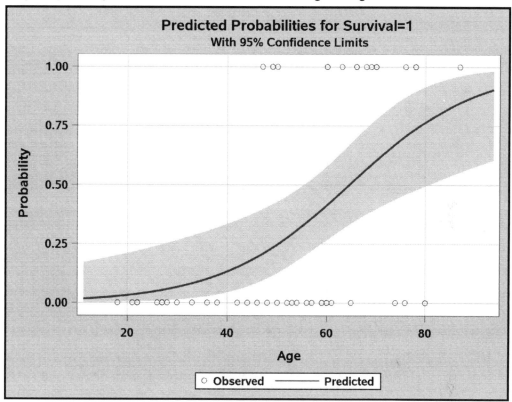

7.2.4 Logistic Regression with All the Explanatory Variables

So finally we can consider a logistic model incorporating all five explanatory variables by amending the previous task as follows:

1. Click the **Binary Logistic Regression** tab.
2. Under **Data ▶ Explanatory Variables ▶ Continuous Variables**, add **shock, malnutrition, alcoholism**, and **bowel_infarc**.
3. Under **Model ▶ Variables**, select all the variables and click **Add**.
4. Click **Run**.

The edited results are shown in Output 7.4. Examining each of the estimated regression coefficients and their corresponding *p*-values, we would conclude that **age, shock**, and **alcoholism** are the important explanatory variables in the prediction of the probability of dying from surgery for intra-abdominal sepsis because the regression coefficients for these variables are all significantly different from 0; aging, the presence of shock, and alcoholism all increase the probability of dying from surgery. One problem with these results is that the relatively small sample size leads to very wide confidence intervals for the associated odds ratios of shock and alcoholism, so we need to be cautious about interpreting the effects of these explanatory variables on survival from surgery. Nevertheless, it is pretty clear that the news for alcoholic patients who suffer shock is not good; their chance of surviving surgery is considerably less than their non-alcoholic and shock-free fellow patients.

We also need to be careful in interpreting each estimated regression coefficient in Output 7.4 in isolation because the explanatory variables are not independent of one another and the same applies to the estimated regression coefficients, a point raised previously in Chapter 6 when we discussed multiple regression models. Removing one of the variables, for example, and refitting the model would lead to different estimated regression coefficients for the variables remaining in the model than for those given in Output 7.4. For this reason, methods for selecting the best set of explanatory variables have been developed when using logistic regression models that are similar to those used in Chapter 6 for multiple regression models, although the criteria used for judging when variables should be removed or entered into an existing model are different. Details are given in Der and Everitt (2013) and exercise 7.5 invites keen readers to apply these variable selection methods for logistic regression to a data set.

Output 7.4: Parameter Estimates and Standard Errors for the Logistic Regression Model with All Five Explanatory Variables Fitted to the Data in Table 7.1

Analysis of Maximum Likelihood Estimates					
Parameter	DF	Estimate	Standard Error	Wald Chi-Square	Pr > ChiSq
Intercept	1	-6.8222	2.2811	8.9448	0.0028
Age	1	0.0731	0.0325	5.0480	0.0247
Shock	1	2.6146	1.1127	5.5219	0.0188
Malnutrition	1	0.7481	0.8909	0.7051	0.4011
Alcoholism	1	2.4143	0.9990	5.8406	0.0157
Bowel_Infarc	1	1.4987	0.9987	2.2518	0.1335

Odds Ratio Estimates		
Effect	Point Estimate	95% Wald Confidence Limits
Age	1.076	1.009 1.147
Shock	13.662	1.543 120.951
Malnutrition	2.113	0.369 12.114
Alcoholism	11.182	1.578 79.228
Bowel_Infarc	4.476	0.632 31.697

7.2.5 A Second Example of the Use of Logistic Regression

Goldberg (1972) describes a psychiatric screening questionnaire, the *General Health Questionnaire* (*GHQ*), designed to identify people who might be suffering from a psychiatric illness. In Table 7.2, some results from applying this instrument are given; here what is of interest is how the probability of being classified as a potential psychiatric case by a psychiatrist is related to an individual's score on the GHQ and the individual's gender. Note that the data here have been grouped in terms of the binary variable, **caseness**, value of yes or no. So, for example, there are four women with a GHQ score of 0 rated as cases and 80 women with a GHQ of 0 rated as non-cases.

Table 7.2: Psychiatric Caseness Data

GHQ Score	Sex	Number of Cases	Number of Non-Cases
0	F	4	80
1	F	4	29

GHQ Score	Sex	Number of Cases	Number of Non-Cases
2	F	8	15
3	F	6	3
4	F	4	2
5	F	6	1
6	F	3	1
7	F	2	0
8	F	3	0
9	F	2	0
10	F	1	0
0	M	1	36
1	M	2	25
2	M	2	8
3	M	1	4
4	M	3	1
5	M	3	1
6	M	2	1
7	M	4	2
8	M	3	1
9	M	2	0
10	M	2	0

7.2.6 An Initial Look at the Caseness Data

A good way to begin to understand the data in Table 7.2 is to plot the estimated probability of being a case against the GHQ score, identifying males and females on the plot.

In addition to the four variables shown in Table 7.2, the data set **sasue.ghq** contains two further variables: **total**, the sum of cases and noncases, and **prcase**, the number of cases divided by the total. We use the **prcase** variable to construct the plot:

1. Open **Tasks ▶ Graph ▶ Scatter Plot**.
2. Under **Data ▶ Data**, select **sasue.ghq**.
3. Under **Data ▶ Roles**, select **score** as the **x variable,**; **prcase** as the **y variable**, and **sex** as the **group variable**.
4. Click **Run**.

Figure 7.2: Plot of Estimated Probability of Being a Case Against the GHQ Score for the Data in Table 7.2

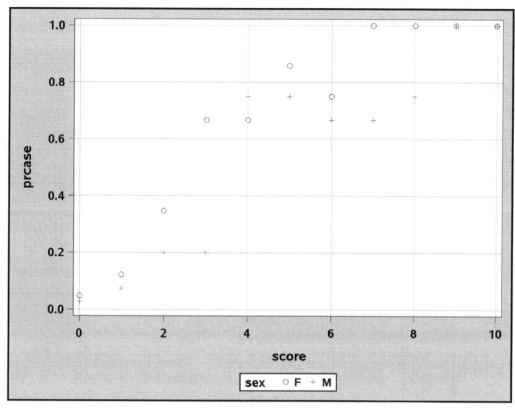

The resulting plot is shown in Figure 7.2. Clearly the estimated probability of being considered a case increases with the increasing GHQ score. For most values of GHQ, women have a higher estimated probability of being considered a case than men, but there is a reversal around a GHQ score of 4. In the next section, we shall fit a logistic model to the caseness data that is suggested by the plot in Figure 7.2.

7.2.7 Modeling the Caseness Data Using Logistic Regression

As suggested by Figure 7.2, we want to fit a model that models the relationship between the probability being considered a case and the GHQ score that allows for both a possible sex difference *and* a possible change in any sex difference over the range of GHQ scores. Such a model has to have explanatory variables for the GHQ score, for the sex, *and* for the *interaction* term score x sex; it is this latter term that allows for the possibility that any sex difference depends on the GHQ score in some way. We can write the required model explicitly as

$$\log\left[\frac{\text{Pr(case)}}{\text{Pr(not case)}}\right] = \beta_0 + \beta_1\text{GHQ} + \beta_2\text{Sex} + \beta_3(\text{GHQxSex})$$

So for men with `sex=1`, the model can be rewritten as

$$\log\left[\frac{\text{Pr(case)}}{\text{Pr(not case)}}\right] = \beta_0 + (\beta_1 + \beta_3)\text{GHQ} + \beta_2$$

And for women with `sex=0`, we have

$$\log\left[\frac{\text{Pr(case)}}{\text{Pr(not case)}}\right] = \beta_0 + \beta_1\text{GHQ}$$

The necessary instructions for fitting this model are:

1. Open **Tasks ▶ Statistics ▶ Binary Logistic Regression**.
2. Under **Data ▶ Data**, select **sasue.ghq**.
3. Under **Data ▶ Roles ▶ Response**, select **Response data consists of number of events and trials**.
4. Add **cases** as the number of events.
5. Add **total** as the number of trials.
6. Under **Data ▶ Roles ▶ Explanatory Variables ▶ Classification Variables**, add **sex** and under **Parameterization of Effects**, select **Reference coding**.
7. Under **Data ▶ Roles ▶ Explanatory Variables ▶ Continuous Variables**, add **score**.
8. Under **Model ▶ Model Effects**, select both variables and click **Full Factorial**. This is equivalent to selecting both and then clicking **Add** and **Cross**.
9. Under **Options ▶ Plots**, select **Default & additional plots** and then select **Effect plot**.
10. Click **Run**.

The parameter estimates and their standard errors are shown in Output 7.5 and the plot representing the fitted model is shown in Figure 7.3.

Output 7.5: Parameter Estimates for the Interaction Model Fitted to the Caseness Data

Parameter		DF	Estimate	Standard Error	Wald Chi-Square	Pr > ChiSq
Intercept		1	-2.9984	0.4926	37.0528	<.0001
score		1	0.6393	0.1225	27.2394	<.0001
sex	F	1	0.2253	0.6093	0.1367	0.7116
score*sex	F	1	0.3020	0.1990	2.3023	0.1292

(Header row: Analysis of Maximum Likelihood Estimates)

Figure 7.3: Plot of Predicted Probability of Caseness from the Interaction Model Fitted to the Caseness Data

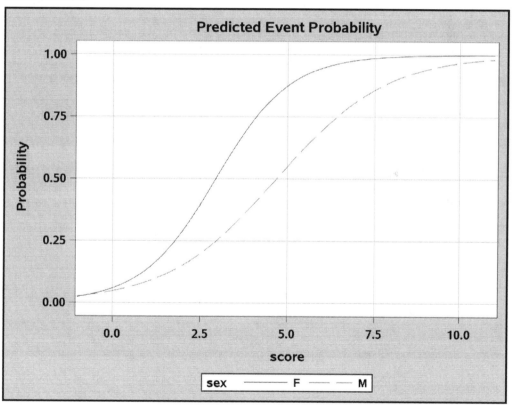

Looking at the parameter estimates in Output 7.5, we see that there is no evidence that the regression coefficients for sex or the interaction of sex and the GHQ score differ from 0. But we should now fit a model that includes both sex and GHQ score, rather than dropping both the interaction of sex with score *and* sex at the same time, because these two effects are confounded. So now let us fit a logistic regression model with **sex** and **score** as the two explanatory variables:

1. Click the **Binary Logistic Regression** tab. (If it has been closed, repeat steps 1 to 8 above.)
2. Under **Model ▶ Model Effects**, delete **score*sex** from the model effects-that is, select it and click the **Delete** button (🗑).
3. Click **Run**.

The estimated regression coefficients for this model are shown in Output 7.6 along with the estimated odds ratio estimate for both the GHQ score and sex and the associated 95% confidence intervals. The model fit statistics have been discussed in the previous chapter and are essentially for use in comparing two competing models (see the next chapter for more details). The Wald chi-squared statistic is simply the square of the ratio of the estimated regression coefficient to its estimated standard error and is tested as a chi-square with a single degree of freedom. Both sex and GHQ score are significant at the 5% level. The confidence intervals show that conditional on sex, a one-point increase in the GHQ score leads to between about an 80% and a 160% increase in the odds of being judged a case rather than a non-case. It also shows that for a given GHQ score, there is between about an 8% and a 500% increase in the odds of being judged a case rather than a non-case for women compared to men. The plot representing the fitted model is shown in Figure 7.4.

Output 7.6: Results from Fitting a Logistic Regression Model with GHQ Score and Sex as Explanatory Variables to the Psychiatric Caseness Data

Analysis of Maximum Likelihood Estimates						
Parameter		DF	Estimate	Standard Error	Wald Chi-Square	Pr > ChiSq
Intercept		1	-3.4296	0.4627	54.9390	<.0001
score		1	0.7791	0.0990	61.8891	<.0001
sex	F	1	0.9361	0.4343	4.6446	0.0312

Odds Ratio Estimates			
Effect	Point Estimate	95% Wald Confidence Limits	
score	2.180	1.795	2.646
sex F vs M	2.550	1.088	5.974

Figure 7.4: Plot of Predicted Probability of Caseness from the Model with Sex and GHQ Score Fitted to the Caseness Data

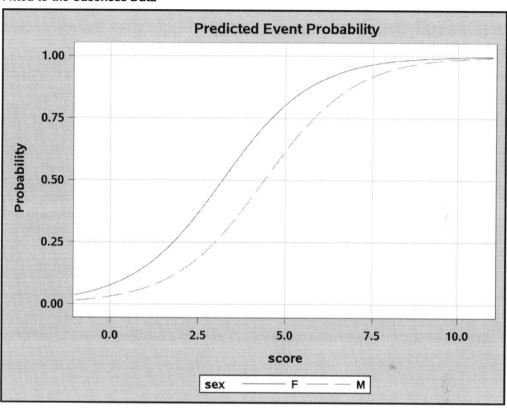

(A more detailed analysis of the caseness data is given in Der and Everitt, 2013.)

7.3 Logistic Regression for 1:1 Matched Studies

A frequently used type of study design, particularly in medical investigations, is known as the *matched case-control* design, in which each person having a particular condition of interest is matched to one or more people without the condition on variables such as age, gender, ethnic group, and so on. A design with *m* controls per case is known as a 1:*m* matched study. In many cases, *m* will be 1, and it is the 1-1 matched study that we shall concentrate on here.

The example we shall consider involves the birth weight of babies. The data arise from looking first at 59 babies who were low weight, defined as weighing less than 2,500 g. The matched data were obtained by randomly selecting for each woman who gave birth to a low birth weight baby, a mother of the same age who did not give birth to a low birth weight baby. Three of the low birth weight mothers were too young to find a match, so the data consist of 56 matched case-control pairs. The complete data are given in Hosmer and Lemeshow (2000), with data for the first five matched pairs shown here in Table 7.3. Variables selected for investigation were prior pre-term delivery (**ptd**, 1=yes, 0=no), smoking status of the mother during pregnancy (**smoke**, 1=yes, 0=no), history of hypertension (**ht**, 1=yes, 0=no), presence of uterine irritability (**ui**, 1=yes, 0=no), and the weight of the mother at the last menstrual period (**lwt**, pounds).

Table 7.3: Part of the Data from the Matched Case-Control Study of Babies by Birth Weight

PAIR	LOW	AGE	LWT	SMOKE	PTD	HT	UI
1	0	14	135	0	0	0	0
1	1	14	101	1	1	0	0
2	0	15	98	0	0	0	0
2	1	15	115	0	0	0	1
3	0	16	95	0	0	0	0
3	1	16	130	0	0	0	0
4	0	17	103	0	0	0	0
4	1	17	130	1	1	0	1
5	0	17	122	1	0	0	0
5	1	17	110	1	0	0	0

LOW   Low Birth Weight

AGE   Age of Mother

LWT   Weight of Mother at Last Menstrual Period

SMOKE   Smoking Status During Pregnancy

PTD   History of Premature Labor

HT   History of Hypertension

UI   Presence of Uterine Irritability

For a single explanatory variable x, for example, the smoking status during pregnancy in the birth weight data, the form of the logistic model used for matched data involves the probability, ϕ, that in matched pair *i*, for a given value of smoking status during pregnancy (yes, or no), the member of the pair is a case. Specifically the model is

$$\operatorname{logit}(\phi) = \alpha_i + \beta x$$

The odds that a subject who smokes during pregnancy $(x = 1)$ is a low birth weight case equals $\exp(\beta)$ times the odds that a subject who does not smoke $(x = 0)$ is a low birth weight case. The α_i terms model the shared values of the *i*th pair on the matching variables.

The model generalizes to the situation where there are *q* explanatory variables as

$$\operatorname{logit}(\phi) = \alpha_i + \beta_1 + x_1 + \cdots \beta_q x_q$$

Typically, one x_i is an explanatory variable of real interest, such as past exposure to smoking in the example above, with the others being used as a form of statistical control in addition to the variables already controlled by virtue of using them to form matched pairs. The problem with the model above is that the number of parameters increases at the same rate as the sample size, with the consequence that maximum likelihood estimation is no longer viable. We can overcome this problem if we regard the parameters α_i as of little interest and so are willing to forgo their estimation. If we do, we can then create a *conditional likelihood function* that will yield maximum conditional likelihood estimators of the coefficients, $\beta_1 \cdots \beta_q$, that are consistent and asymptotically normally distributed. Essentially, the conditioning avoids the estimation of the parameters accounting for the matching. The mathematics behind this are described in Collett (2003), but the parameters for the explanatory variables in the model have the same log odds ratio interpretation familiar from the standard logistic model The result is that we can conduct the regression analyses exactly as before. However, the variables used in matching are controlled for automatically and so not used directly in modeling. Here we concentrate on how to fit such a model to the low birth weight data and how to interpret the resulting parameter estimates. We begin with a simple model that only considers the smoking status during pregnancy:

1. Open **Tasks ▶ Statistics ▶ Binary Logistic Regression**.
2. Under **Data ▶ Data**, add **sasue.lbwpairs**.
3. Under **Data ▶ Roles ▶ Response**, add **low** and select the **Event of interest** to be 1.
4. Under **Data ▶ Roles ▶ Explanatory Variable ▶ Continuous Variables**, add **smoke**.
5. Under **Model ▶ Model Effects**, select **smoke** and click **Add**.

    We now need to edit the code produced so far.

6. In the code pane, click **Edit**.

   A program pane opens that contains the following code:

   ```
   proc logistic data=SASUE.LBWPAIRS plots;
       model low(event='1')=Smoke / link=logit technique=fisher;
   run;
   ```

7. Before the RUN statement, add

   ```
   strata pair;
   ```

   Take care to include the semicolon at the end.

8. Click **Run**.

The results are shown in Output 7.7. The 95% confidence interval for the odds ratio is [1.224, 6.177]. This implies that in a matched pair of women, the odds that the mother of a low birthweight baby smoked during pregnancy, are estimated to be between about 1.2 and 6.2 times the odds that the control mother, one not having a low birthweight baby, has smoked during pregnancy.

Output 7.7: Parameter Estimates for the Logistic Regression Model Applied to the Matched Low Birth Weight Data Using Smoking During Pregnancy as the Single Explanatory Variable

Analysis of Conditional Maximum Likelihood Estimates					
Parameter	DF	Estimate	Standard Error	Wald Chi-Square	Pr > ChiSq
Smoke	1	1.0116	0.4129	6.0036	0.0143

Odds Ratio Estimates		
Effect	Point Estimate	95% Wald Confidence Limits
Smoke	2.750	1.224 6.177

We now fit the model with all the explanatory variables by further edits to the code produced above:

1. Select the program window containing the edited code.

2. Type in the names of the other explanatory variables (**lwt**, **ptd**, **ht**, and **ui**) on the MODEL statement so that the resulting code now is

   ```
   proc logistic data=SASUE.LBWPAIRS;
       model low(event='1')=Smoke lwt ptd ht ui  / link=logit
   ```

```
    technique=fisher;
        strata pair;
    run;
```

3. Click **Run**.

The results are shown in Output 7.8.

Output 7.8: Results from Fitting Logistic Regression to the Matched Birth Weight Data Using All Variables

Model Information	
Data Set	SASUE.LBWPAIRS
Response Variable	low
Number of Response Levels	2
Number of Strata	56
Model	binary logit
Optimization Technique	Newton-Raphson ridge

Number of Observations Read	112
Number of Observations Used	112

Response Profile		
Ordered Value	low	Total Frequency
1	0	56
2	1	56

Probability modeled is low=1.

Strata Summary				
Response Pattern	low 0	1	Number of Strata	Frequency
1	1	1	56	112

Newton-Raphson Ridge Optimization

Without Parameter Scaling

Convergence criterion (GCONV=1E-8) satisfied.

Model Fit Statistics		
Criterion	Without Covariates	With Covariates
AIC	77.632	62.474
SC	77.632	76.066
-2 Log L	77.632	52.474

Testing Global Null Hypothesis: BETA=0			
Test	Chi-Square	DF	Pr > ChiSq
Likelihood Ratio	25.1587	5	0.0001
Score	19.7845	5	0.0014
Wald	12.5938	5	0.0275

Analysis of Conditional Maximum Likelihood Estimates					
Parameter	DF	Estimate	Standard Error	Wald Chi-Square	Pr > ChiSq
Smoke	1	1.4796	0.5620	6.9305	0.0085
LWT	1	-0.0151	0.00815	3.4281	0.0641
PTD	1	1.6706	0.7468	5.0041	0.0253
Ht	1	2.3294	1.0025	5.3984	0.0202
UI	1	1.3449	0.6938	3.7571	0.0526

Odds Ratio Estimates			
Effect	Point Estimate	95% Wald Confidence Limits	
Smoke	4.391	1.459	13.212
LWT	0.985	0.969	1.001
PTD	5.315	1.230	22.973

Odds Ratio Estimates		
Effect	Point Estimate	95% Wald Confidence Limits
Ht	10.271	1.440 73.283
UI	3.838	0.985 14.951

The estimated odds ratios in Output 7.8 indicate that smoking during pregnancy, prior preterm deliveries, and the presence of hypertension are important risk factors for delivering a low birth weight baby. The confidence intervals associated with the dichotomous variables are very wide, which largely results from the sparsity of discordant pairs. Hosmer and Lemeshow (2005) point out that the gain in precision obtained from matching and using conditional logistic regression may be offset by a loss owing to there being only a few discordant pairs for dichotomous covariates.

7.4 Summary

Logistic regression models are very widely used in many disciplines but particularly in medical studies. In this chapter, we have tried to describe how to interpret the estimated parameters from such models. As with multiple regression models, a complete analysis of a data set using logistic regression should involve diagnostic plots for assessing the assumptions made by these models. Such diagnostic plots are provided by SAS University Edition, but we have not discussed them here because they are technically quite complicated and, in our experience, they are not always helpful in shedding light on whether a model is suitable for a particular data set. Full details of diagnostic plots for logistic regression models are, however, given in Collett (2003) for readers who are anxious that they may be missing something of importance.

7.5 Exercises

Exercise 7.1: Plasma Data

The data set **plasma** was collected to examine the extent to which erythrocyte sedimentation rate (ESR), the rate at which red blood cells (erythocytes) settle out of suspension in blood plasma, is related to two plasma proteins, fibrinogen and γ-globulin, both measured in gm/l. The ESR for a healthy individual should be less than 20mm/h and because the absolute value of ESR is relatively unimportant, the response variable used here denotes whether this is the case. A response of 0 signifies a healthy individual (ESR<20), while a response of unity refers to an unhealthy individual (ESR≥20). The aim of the analysis for these data is to determine the strength of any relationship between the ESR level and the levels of the two plasmas. Investigate the relationship by fitting a logistic model for the probability of an unhealthy individual with **fibrinogen** and **gamma** as the two explanatory variables. What are your conclusions?

Fibrinogen	Gamma	ESR
2.52	38	0
2.56	31	0
2.19	33	0
2.18	31	0
3.41	37	0
2.46	36	0
3.22	38	0
2.21	37	0
3.15	39	0
2.60	41	0
2.29	36	0
2.35	29	0
5.06	37	1
3.34	32	1
2.38	37	1
3.15	36	0
3.53	46	1
2.68	34	0
2.60	38	0
2.23	37	0
2.88	30	0
2.65	46	0
2.09	44	1
2.28	36	0
2.67	39	0
2.29	31	0
2.15	31	0
2.54	28	0
3.93	32	1
3.34	30	0
2.99	36	0
3.32	35	0

Exercise 7.2: Leukaemia Data

The data in **leukaemia2** report whether patients with leukaemia lived for at least 24 weeks after diagnosis, along with the values of two explanatory variables, the white blood count and the presence or absence of a morphological characteristic of the white blood cells (AG). The data are from Venables and Ripley (1994). Fit a logistic regression model to the data to determine whether the explanatory variables are predictive of survival longer than 24 weeks. (You may need to consider an interaction term.)

White Blood Count	AG	Survival Longer than 24 Weeks from Diagnosis (1=yes, 0=no)
2300	present	1
750	present	1
4300	present	1
2600	present	1
6000	present	0
10500	present	1
10000	present	1
17000	present	0
5400	present	1
7000	present	1
9400	present	1
32000	present	1
35000	present	0
100000	present	0
100000	present	0
52000	present	0
100000	present	1
4400	absent	1
3000	absent	1
4000	absent	0
1500	absent	0
9000	absent	0
5300	absent	0
10000	absent	0

White Blood Count	AG	Survival Longer than 24 Weeks from Diagnosis (1=yes, 0=no)
19000	absent	0
27000	absent	0
28000	absent	0
31000	absent	0
26000	absent	0
21000	absent	0
79000	absent	1
100000	absent	0
100000	absent	1

Exercise 7.3: Low Birth Weight Data

The data in **lowbwgt** comprise part of the data set given in Hosmer and Lemeshow (1989), collected during a study to identify risk factors associated with giving birth to a low birth weight baby, defined as weighing less than 2,500 grams. The risk factors considered were the age of the mother, the weight of the mother at her last menstrual period, the race of the mother, and the number of physician visits during the first trimester of the pregnancy. Fit a logistic regression model for the probability of a low birth weight infant using **age**, **lwt**, **race** (coded in terms of two dummy variables), and **ftv** as explanatory variables. What conclusions do you draw from the fitted model?

ID	LOW	AGE	LWT	RACE	FTV
85	0	19	182	2	0
86	0	33	155	3	3
87	0	20	105	1	1
88	0	21	108	1	2
. . . .					
79	1	28	95	1	2
81	1	14	100	3	2
82	1	23	94	3	0
83	1	17	142	2	0
84	1	21	130	1	3

LOW	:	0 = weight of baby > 2,500 g
		1 = weight of baby <= 2,500 g
AGE	:	Age of mother in years
LWT	:	Weight of mother at last menstrual period
RACE	:	1 = white, 2 = black, 3 = other
FTV	:	Number of physician visits in the first trimester

Exercise 7.4: Role Data

In a survey carried out in 1974 and 1975, each respondent was asked if he or she agreed or disagreed with the following statement: Women should take care of running their homes and leave running the country to men. The responses are summarized in the **role** data set (from Haberman, 1973) and are also given in Collett (2003). The questions of interest here are whether the responses of men and women differ and how years of education affect the response given. Fit appropriate logistic regression models to answer these questions.

Years of Education	Gender	Agree	Disagree
0	M	4	2
1	M	2	0
2	M	4	0
3	M	6	3
4	M	5	5
5	M	13	7
6	M	25	9
7	M	27	15
8	M	75	49
9	M	29	29
10	M	32	45
11	M	36	59
12	M	115	245
13	M	31	70
14	M	28	79

Years of Education	Gender	Agree	Disagree
15	M	9	23
16	M	15	110
17	M	3	29
18	M	1	28
19	M	2	13
20	M	2	20
0	F	4	2
1	F	1	0
2	F	0	0
3	F	6	1
4	F	10	0
5	F	14	7
6	F	17	5
7	F	26	16
8	F	91	36
9	F	30	35
10	F	55	67
11	F	50	62
12	F	190	403
13	F	17	92
14	F	18	81
15	F	7	34
16	F	13	115
17	F	3	28
18	F	0	21
19	F	1	2
20	F	2	4

Exercise 7.5: Backward Elimination

Investigate the use of backward elimination in the context of logistic regression modelling to try to find a more parsimonious model for the low birth weight data in Table 7.14.

Exercise 7.6: Acute Herniated Lumber Disc Data

Kelsey and Hardy (1975) describe a study designed to investigate whether driving a car is a risk factor for lower back pain resulting from acute herniated lumber invertebral discs (AHLID). A case-control study was used with cases selected from people who had recently had X-rays taken of the lower back and who had been diagnosed as having AHLID. The controls were taken from patients admitted to the same hospital as a case with a condition unrelated to the spine. Further matching was made on age and sex and a total of 217 matched pairs were recruited, consisting of 89 female pairs and 128 male pairs. The complete data are available in the **ahlid** data set. Only a part is shown here:

Pair	Status	Driver	Suburban Resident
1	case	1	1
1	control	1	0
2	case	1	1
2	control	1	1
3	case	1	0
3	control	1	1
. . .			
216	case	0	0
216	control	1	1
217	case	1	1
217	control	1	0

Fit a conditional logistic regression model to the data to determine whether driving and area of residence are risk factors for AHLID.

Chapter 8: Poisson Regression and the Generalized Linear Model

---

8.1 Introduction

In this chapter, we show how the multiple linear regression model and the logistic regression model can both be subsumed within a more general modelling framework, the *generalized linear model* (GLM), and how the GLM then leads to more appropriate models for response variables that cannot (and should not) be modelled by the methods described in either Chapter 6 or Chapter 7. The statistical topics to be covered are

- Generalized linear model

- Linear and logistic regression fitted using the generalized linear model

- Poisson regression

8.2 Generalized Linear Model

The term *generalized linear model* (*GLM*) was first introduced in a landmark paper by Nelder and Wedderburn (1972), in which a wide range of seemingly disparate problems of statistical modeling and inference were set in an elegant, unifying framework of great power and flexibility. Generalized linear models include all the modeling techniques described in earlier chapters (that is analysis of variance, analysis of covariance, multiple linear regression, and logistic regression) and open up the possibility of other models--for example, *Poisson regression*, which we shall describe in this chapter. A comprehensive account of GLMs is given in McCullagh and Nelder (1989) and a more concise and less technical description in Dobson and Barnett (2008). In the next subsection, we very briefly review the main features of such models.

8.2.1 Components of Generalized Linear Models

The multiple linear regression model described in Chapter 6 has the form

$$y = \beta_0 + \beta_1 x_1 \cdots + \beta_q x_q + \varepsilon$$

The error term, ε, is assumed to have a normal distribution with zero mean and constant variance, σ^2. An equivalent way of writing the model is as $y \sim N(\mu, \sigma^2)$ where $\mu = \beta_0 + \beta_1 x_1 + ... + \beta_q x_q$. This makes it clear that this model is only suitable for continuous response variables with, conditional on the values of the explanatory variables, a normal distribution with constant variance. The generalization of such a model made in GLMs consists of allowing each of the following three assumptions associated with the multiple linear regression model to be modified:

- The response variable is normally distributed with a mean determined by the model.
- The mean can be modelled as a linear function (possibly nonlinear transformations) of the explanatory variables (that is, the effects of the explanatory variables on the mean are additive).
- The variance of the response variable, given the (predicted) mean, is constant.

In a GLM, some *transformation* of the mean is modelled by a linear function of the explanatory variables, and the distribution of the response variable around its mean (often referred to as the *error distribution*) is generalized, usually in a way that fits naturally with a particular transformation. (We shall deal with the last bullet point above later in the chapter.) The result is a very wide class of regression models that includes many other models as special cases (for example, analysis of variance, multiple linear regression, and logistic regression). The three essential components of a GLM are

1. A linear predictor, η , formed from the explanatory variables:

$$\eta = \beta_0 + \beta_1 x_1 + \beta_2 x_2 \cdots + \beta_q x_q$$

2. A transformation of the mean, μ , of the response variable called the *link function*, $g(\mu)$. In a GLM, it is $g(\mu)$, which is modelled by the linear predictor:

$$g(\mu) = \eta$$

In multiple linear regression and analysis of variance, the link function is the identity function so that the mean is modelled directly as a linear function of the explanatory variables. In logistic regression, the link function is the logit--that is, the log of the odds ratio is modelled.

3. The distribution of the response variable given its mean μ is assumed to be of some particular type from a family of distributions known as the *exponential family*; for details, see Dobson and Barnett (2008).

    The parameters in GLMs are estimated by maximizing the joint likelihood of the observed responses given the parameters of the model and the explanatory variables. This generally requires iterative numerical algorithms. See McCullagh and Nelder (1989) for details.

8.2.2 Using the Generalized Linear Model to Apply Multiple Linear Regression and Logistic Regression

Here we will illustrate how two sets of data, the ice cream consumption data used in Chapter 6 and the survival from surgery data that appeared in Chapter 7, can be analysed by applying the same models used in those two previous chapters within a generalized linear modelling framework.

So, first the ice cream consumption data where the SAS University Edition code now is:

1. Open **Tasks ▶ Statistics ▶ Generalized Linear Models**.
2. Under **Data ▶ Data**, select **sasue.icecream**.
3. Under **Data ▶ Roles ▶ Response ▶ Distribution**, select **normal** and add **consumption** as the **response variable**.
4. Under **Data ▶ Roles ▶ Explanatory Variables ▶ Continuous Variables**, add **price** and **temperature**.
5. Under **Model ▶ Model Effects**, add **price** and **temperature**.
6. Click **Run**.

The edited output is shown in Output 8.1. The first part of the output identifies the data set, the distribution assumed, and the link function. The next part lists the values of a number of goodness-of-fit indicators. Some of these, such as the AIC and AICC, have been briefly discussed in Chapter 6; the BIC, standing for Bayesian information criterion, is similar to the AIC, but penalizes models of high dimensionality more than that index. The deviance we will explain later in the chapter. The parameter estimates are found next, and these are the same as those given in Chapter 6. The scale parameter mentioned will be explained in Section 8.4. The Wald statistic is simply the ratio of an estimated parameter to its estimated standard error. Under the hypothesis that the population parameter value is 0, the square of the Wald statistic has a chi-square distribution with a single degree of freedom.

Output 8.1: Edited Output from Applying Generalized Linear Modelling to the Ice Cream Consumption Data

Model Information	
Data Set	SASUE.ICECREAM
Distribution	Normal
Link Function	Identity
Dependent Variable	consumption

Number of Observations Read	30
Number of Observations Used	30

Criteria For Assessing Goodness Of Fit			
Criterion	DF	Value	Value/DF
Deviance	27	0.0461	0.0017
Scaled Deviance	27	30.0000	1.1111
Pearson Chi-Square	27	0.0461	0.0017
Scaled Pearson X2	27	30.0000	1.1111
Log Likelihood		54.6072	
Full Log Likelihood		54.6072	
AIC (smaller is better)		-101.2144	
AICC (smaller is better)		-99.6144	
BIC (smaller is better)		-95.6096	

Algorithm converged.

Analysis Of Maximum Likelihood Parameter Estimates							
Parameter	DF	Estimate	Standard Error	Wald 95% Confidence Limits		Wald Chi-Square	Pr > ChiSq
Intercept	1	0.5966	0.2451	0.1163	1.0768	5.93	0.0149
price	1	-1.4018	0.8776	-3.1219	0.3183	2.55	0.1102
temperature	1	0.0030	0.0004	0.0022	0.0039	46.20	<.0001
Scale	1	0.0392	0.0051	0.0304	0.0505		

Note: The scale parameter was estimated by maximum likelihood.

Now we can move on to the analysis of the survival for surgery data, where the logistic regression model described in Chapter 7 can be fitted within the generalized linear model framework as follows:

1. Open **Tasks ▸ Statistics ▸ Generalized Linear Models**.
2. Under **Data ▸ Data**, select **sasue.sepsis**.
3. Under **Data ▸ Roles ▸ Response ▸ Distribution**, select **Binomial**, add **survival** as the **response** variable, and select **Higher value** as the **event of interest**.
4. Under **Data ▸ Roles ▸ Explanatory Variables ▸ Continuous Variables**, add all the remaining variables except **ID**.
5. Under **Model ▸ Model Effects**, add all the variables.
6. Click **Run**.

The edited output is shown in Output 8.2. Here the first part of the output details that the distribution assumption is binomial and the link is the logit function, giving the logistic regression model. The parameter estimates are the same as those given in Output 7.4 of Chapter 7. Again, the scale function mentioned will be discussed later.

Output 8.2: Edited Output from Applying the Generalized Linear Model to the Survival from Surgery Data

Model Information	
Data Set	SASUE.SEPSIS
Distribution	Binomial
Link Function	Logit
Dependent Variable	Survival

Number of Observations Read	50
Number of Observations Used	50
Number of Events	18
Number of Trials	50

Response Profile		
Ordered Value	Survival	Total Frequency
1	1	18
2	0	32

PROC GENMOD is modeling the probability that Survival='1'.

Criteria For Assessing Goodness Of Fit			
Criterion	DF	Value	Value/DF
Log Likelihood		-18.0381	
Full Log Likelihood		-18.0381	
AIC (smaller is better)		48.0762	
AICC (smaller is better)		50.0297	
BIC (smaller is better)		59.5483	

Algorithm converged.

Analysis Of Maximum Likelihood Parameter Estimates							
Parameter	DF	Estimate	Standard Error	Wald 95% Confidence Limits		Wald Chi-Square	Pr > ChiSq
Intercept	1	-6.8222	2.2811	-11.2930	-2.3514	8.94	0.0028
Shock	1	2.6146	1.1127	0.4338	4.7954	5.52	0.0188
Malnutrition	1	0.7481	0.8909	-0.9981	2.4943	0.71	0.4011
Alcoholism	1	2.4143	0.9990	0.4563	4.3723	5.84	0.0157
Age	1	0.0731	0.0325	0.0093	0.1369	5.05	0.0247

Analysis Of Maximum Likelihood Parameter Estimates							
Parameter	DF	Estimate	Standard Error	Wald 95% Confidence Limits		Wald Chi-Square	Pr > ChiSq
Bowel_Infarc	1	1.4987	0.9987	-0.4588	3.4562	2.25	0.1335
Scale	0	1.0000	0.0000	1.0000	1.0000		

Note: The scale parameter was held fixed.

8.3 Poisson Regression

The Poisson regression model is useful for a response variable, *y*, that is a count or frequency and for which it is reasonable to assume an underlying Poisson distribution, that is:

$$\Pr(y) = \frac{\mu^y e^{-\mu}}{y!} \quad y = 0, 1, 2 \ldots$$

In Poisson regression, it is the log of the response variable's mean that is modelled as a linear function of the explanatory variables. In other words, in the GLM framework, the link function is the log function and the error distribution is Poisson. The log link is needed to prevent any predicted values of the response variable being negative.

8.3.1 Example 1: Familial Adenomatous Polyposis (FAP)

As an example of Poisson regression, we shall apply the method to the data shown in Table 8.1 taken from Piantadosi (1997). The data arise from a study of Familial Adenomatous Polyposis (FAP), an autosomal dominant genetic defect that predisposes those affected to develop large numbers of polyps in the colon, which, if untreated, could develop into colon cancer. Patients with FAP were randomly assigned to receive an active drug treatment or a placebo. The response variable was the number of colonic polyps at 3 months after starting treatment. Additional covariates of interest were the number of polyps before starting treatment, gender, and age.

Table 8.1: FAP Data

Sex	Treatment	Baseline Count of Polyps	Age	Number of Polyps at 3 Months
0	1	7	17	6
0	0	77	20	67
1	1	7	16	4

Sex	Treatment	Baseline Count of Polyps	Age	Number of Polyps at 3 Months
0	0	5	18	5
1	1	23	22	16
0	0	35	13	31
0	1	11	23	6
1	0	12	34	20
1	0	7	50	7
1	0	318	19	347
1	1	160	17	142
0	1	8	23	1
1	0	20	22	16
1	0	11	30	20
1	0	24	27	26
1	1	34	23	27
0	0	54	22	45
1	1	16	13	10
1	0	30	34	30
0	1	10	23	6
0	1	20	22	5
1	1	12	42	8

Sex: 0=female, 1=male

Treatment: 0=placebo, 1=active

A Poisson regression model can be fitted as follows:

1. Open **Tasks ▶ Statistics ▶ Generalized Linear Models**.
2. Under **Data ▶ Data**, select **sasue.fap**.
3. Under **Data ▶ Roles ▶ Response ▶ Distribution**, select **Poisson** and add **Nat3mth** as the response variable.
4. Under **Data ▶ Roles ▶ Explanatory Variables ▶ Continuous Variables**, add all the remaining variables.
5. Under **Model ▶ Model Effects**, add all the variables.
6. Click **Run**.

The numerical results are shown in Output 8.3. The regression coefficients become easier to interpret if they (and the confidence limits) are exponentiated. For example, the exponentiated confidence interval for the gender regression coefficient is [1.066, 1.647]; men are estimated to have somewhere between about 7% and 65% more polyps at 3 months than women, conditional on the other covariates being the same. For treatment, the corresponding interval is [0.600, 0.882]; patients receiving the active treatment are estimated to have between 60% and 88% of the number of polyps at 3 months than those receiving the placebo, again conditional on the other covariates being equal.

Here we should say something about the deviance term in Output 8.3 and previous output in this chapter. This is essentially a measure of how a model fits the data and is most useful in comparing two competing *nested* models (that is, two models in which one contains terms that are additional to those in the other). The difference in deviance between two nested models measures the extent to which the additional terms improve the fit of the more complex model. The difference in deviance has approximately a chi-square distribution with degrees of freedom equal to the difference in the degrees of freedom of the two nested models. Exercise 8.3 gives an opportunity for readers to investigate using the deviance in this way.

(For the multiple regression model, a generalized linear model with a normal distribution and an identity link function, the deviance is simply the residual sum of squares from the usual analysis of variance table for such a model. See Chapter 6 for more information.)

It should be noted that one aspect of the fitted Poisson regression model to the FAP data, namely the value of the deviance divided by the degrees of freedom (the value 10.98 in Output 8.3) has implications for the appropriateness of the model, a point we shall take up in detail in the next section.

Output 8.3: Numerical Output from Fitting Poisson Regression to the FAP Data

Model Information	
Data Set	SASUE.FAP
Distribution	Poisson
Link Function	Log
Dependent Variable	Nat3mth

Number of Observations Read	22
Number of Observations Used	22

Criteria For Assessing Goodness Of Fit			
Criterion	DF	Value	Value/DF
Deviance	17	186.7304	10.9841
Scaled Deviance	17	186.7304	10.9841
Pearson Chi-Square	17	186.0802	10.9459
Scaled Pearson X2	17	186.0802	10.9459
Log Likelihood		2946.0059	
Full Log Likelihood		-143.8490	
AIC (smaller is better)		297.6980	
AICC (smaller is better)		301.4480	
BIC (smaller is better)		303.1533	

Algorithm converged.

Analysis Of Maximum Likelihood Parameter Estimates							
Parameter	DF	Estimate	Standard Error	Wald 95% Confidence Limits		Wald Chi-Square	Pr > ChiSq
Intercept	1	3.3610	0.1882	2.9922	3.7298	319.09	<.0001
sex	1	0.2814	0.1111	0.0637	0.4991	6.42	0.0113
treatment	1	-0.3183	0.0984	-0.5112	-0.1254	10.46	0.0012
baseline_n	1	0.0089	0.0004	0.0081	0.0097	479.32	<.0001
age	1	-0.0264	0.0073	-0.0408	-0.0120	12.95	0.0003
Scale	0	1.0000	0.0000	1.0000	1.0000		

Note: The scale parameter was held fixed.

8.3.2 Example 2: Bladder Cancer

As a further example of the use of Poisson regression, we shall use the data shown in Table 8.2, taken from Seeber (1989). The data arise from 31 male patients who have been treated for superficial bladder cancer. They give the number of recurrent tumours during a particular time period after removal of the primary tumour and the size of the primary tumour (whether smaller or larger than 3 cm).

Table 8.2: Bladder Cancer Data (Used with Permission of the Publishers, John Wiley and Sons Ltd, from Seeber, 1998)

Time	X	n
2	0	1
3	0	1
6	0	1
8	0	1
9	0	1
10	0	1
11	0	1
13	0	1
14	0	1
16	0	1
21	0	1
22	0	1
24	0	1
26	0	1
27	0	1
7	0	2
13	0	2
15	0	2
18	0	2
23	0	2
20	0	3
24	0	4
1	1	1
5	1	1
17	1	1
18	1	1
25	1	1
18	1	2
25	1	2
4	1	3
19	1	4

X = 0 tumour < 3 cm

X = 1 tumour > 3 cm

Before coming to the analysis of the data in Table 8.2, we first need to introduce the idea of a *Poisson process*, in which the waiting times between successive events of interest (the tumours, in this case) are independent and exponentially distributed with a common mean, $1 \backslash \lambda$ (say). Then, the number of events that occur up to time t has a Poisson distribution with mean $\mu = \lambda t$. Here the parameter of real interest is the rate at which events occur, λ, and for a single explanatory variable, x, we can adopt a Poisson regression approach using the following model to examine the dependence of λ on x:

$$\log \lambda = \log \frac{\mu}{t} = \beta_0 + \beta_1 x$$

Rearranging this model, we obtain

$$\log \mu = \beta_0 + \beta_1 x + \log t$$

In this form, the model can be fitted within the GLM framework with a log link and Poisson errors. But in this model $\log t$ is a variable in the model whose regression coefficient is *fixed* at unity and is usually known as an *offset*. To create the offset variable, we first transform the time variable to log_time using the Transform Data task as follows:

1. Open **Tasks ▶ Data ▶ Transform Data**.
2. Under **Data ▶ Data**, select **sasue.bladder**.
3. Under **Data ▶ Transform 1 ▶ Variable 1**, add **time** and select **Natural Log** as the transform.
4. Under **Data ▶ Output Data Set**, delete **Transform** and type **sasue.bladder**.
5. Click **Run**.

The Poisson regression model is fitted to the bladder cancer data as follows:

1. Open **Tasks ▶ Statistics ▶ Generalized Linear Models**.
2. Under **Data ▶ Data**, select **sasue.bladder**.
3. Under **Data ▶ Roles ▶ Response ▶ Distribution**, select **Poisson** and add **n** as the **response** variable.
4. Under **Data ▶ Roles ▶ Explanatory Variables ▶ Continuous** Variables, add **x**.
5. Under **Data ▶ Roles ▶ Explanatory Variables ▶ Offset** Variable, add **log_time**.

6. Under **Model ▶ Model Effects**, add **x.** The offset, **log_time**, is added automatically.
7. Click **Run.**

The numerical results are shown in Output 8.4. The estimated model is

$$\log \lambda = -2.339 + 0.229x$$

So, for smaller tumors (x=0), the estimated (baseline) rate is exp(-2.339)=0.096, and for larger tumors (x=1), the estimated rate is exp(-2.339+0.229)=0.12. The rate for larger tumors is estimated as 0.12/0.096=1.25 times the rate for smaller tumours. In terms of waiting times between recurrences, the means are 1/0.096=10.42 months for smaller tumours and 1/0.12=8.33 months for larger tumors. But the regression coefficient for the dummy variable that codes for tumour size is seen from Output 8.4 to be non-significant, so the data give no evidence that rates or waiting times for large and small tumours are different. This becomes apparent if we construct a confidence interval for the rate for larger tumours from the confidence limits given in Output 8.4 as [exp(-2.339-0.371), exp(-2.339+0.829)], that is [0.067, 0.221]. This interval contains the rate for smaller tumours. There is no evidence that the size of the primary tumour is associated with the number of recurrent tumours.

Output 8.4: Numerical Results from Fitting the Poisson Regression Model to the Bladder Cancer Data

Model Information	
Data Set	SASUE.BLADDER
Distribution	Poisson
Link Function	Log
Dependent Variable	n
Offset Variable	log_time

Number of Observations Read	31
Number of Observations Used	31

Criteria For Assessing Goodness Of Fit			
Criterion	DF	Value	Value/DF
Deviance	29	25.4189	0.8765
Scaled Deviance	29	25.4189	0.8765
Pearson Chi-Square	29	38.5938	1.3308

Criteria For Assessing Goodness Of Fit			
Criterion	DF	Value	Value/DF
Scaled Pearson X2	29	38.5938	1.3308
Log Likelihood		-33.3234	
Full Log Likelihood		-48.1150	
AIC (smaller is better)		100.2301	
AICC (smaller is better)		100.6586	
BIC (smaller is better)		103.0980	

Algorithm converged.

Analysis Of Maximum Likelihood Parameter Estimates							
Parameter	DF	Estimate	Standard Error	Wald 95% Confidence Limits		Wald Chi-Square	Pr > ChiSq
Intercept	1	-2.3394	0.1768	-2.6859	-1.9929	175.13	<.0001
x	1	0.2292	0.3062	-0.3709	0.8293	0.56	0.4541
Scale	0	1.0000	0.0000	1.0000	1.0000		

Note: The scale parameter was held fixed.

8.4 Overdispersion

An important aspect of generalized linear models that thus far we have largely ignored is the variance function, $V(\mu)$, that captures how the variance of a response variable depends on its mean. The general form of the relationship is $\text{Var}(\text{response}) = \phi V(\mu)$, where ϕ is a constant and $V(\mu)$ specifies how the variance depends on the mean, μ. For the error distributions considered previously, this general form becomes

1. Normal: $V(\mu) = 1, \phi = \sigma^2$; here the variance does not depend on the mean.

2. Binomial: $V(\mu) = \mu(1 - \mu), \phi = 1$;

3. Poisson: $V(\mu) = \mu; \phi = 1$

In the case of a Poisson variable, we see that the mean and variance are equal, and in the case of a binomial variable where the mean is the probability of the occurrence of the event of interest, p, the variance is $p(1 - p)$.

Both the Poisson and binomial distributions have variance functions that are completely determined by the mean. There is no free parameter for the variance because in applications of the generalized linear model with binomial or Poisson error distributions, the dispersion parameter, ϕ, is defined as 1 (see previous results for logistic and Poisson regression). But in some applications, this becomes too restrictive to fully account for the empirical variance in the data; in such cases, it is common to describe the phenomenon as *overdispersion*. For example, if the response variable is the proportion of family members who have been ill in the past year, observed in a large number of families, then the individual binary observations that make up the observed proportions are likely to be correlated rather than independent. This non-independence can lead to a variance that is greater (less) than that on the assumption of binomial variability. And observed counts often exhibit larger variance than would be expected from the Poisson assumption, a fact noted by Greenwood and Yule over 80 years ago (Greenwood and Yule, 1920). Greenwood and Yule's suggested solution to the problem was a model in which μ was a random variable with a gamma distribution leading to a *negative binomial distribution* for the count.

There are a number of strategies for accommodating overdispersion, but here we concentrate on a relatively simple approach that retains the use of the binomial or Poisson error distributions as appropriate but allows estimation of a value of ϕ from the data rather than defining it to be unity for these distributions. The estimate is usually the residual deviance divided by its degrees of freedom, exactly the method used with Gaussian models. Parameter estimates remain the same, but parameter standard errors are increased by multiplying them by the square root of the estimated dispersion parameter. This process can be carried out manually, or almost equivalently the overdispersed model can be formally fitted using a procedure known as *quasi-likelihood*. This allows estimation of model parameters without fully knowing the error distribution of the response variable. See McCullagh and Nelder (1989) for full technical details on this approach.

When fitting generalized linear models with binomial or Poisson error distributions, overdisperson can often be spotted by comparing the residual deviance with its degrees of freedom. For a well-fitting model, the two quantities should be approximately equal. If the deviance is far greater than the degrees of freedom, overdispersion might be indicated. In Output 8.3, for example, we see that the ratio of deviance to degrees of freedom is nearly 11, clearly indicating an overdispersion problem. Consequently, we will now refit the Poisson model with the square root of the deviance divided by its degrees of freedom as the scale parameter using the following codeL

1. Open **Tasks ▶ Statistics ▶ Generalized Linear Models**.
2. Under **Data ▶ Data**, select **sasue.fap**.
3. Under **Data ▶ Roles ▶ Response ▶ Distribution**, select **Poisson** and add **Nat3mth** as the **response** variable.

4. Under **Data ▶ Roles ▶ Explanatory Variables ▶ Continuous Variables**, add all the remaining variables.
5. Under **Model ▶ Model Effects**, add all the variables.
6. Under **Options ▶ Methods ▶ Dispersion**, select **Adjust for dispersion** and from the drop-down menu, select **Use deviance estimate**.
7. Click **Run**.

Output 8.5: Numeric Results of Fitting Overdispersed Poisson Model to FAP Data

Model Information	
Data Set	SASUE.FAP
Distribution	Poisson
Link Function	Log
Dependent Variable	Nat3mth

Number of Observations Read	22
Number of Observations Used	22

Criteria For Assessing Goodness Of Fit			
Criterion	DF	Value	Value/DF
Deviance	17	186.7304	10.9841
Scaled Deviance	17	17.0000	1.0000
Pearson Chi-Square	17	186.0802	10.9459
Scaled Pearson X2	17	16.9408	0.9965
Log Likelihood		268.2054	
Full Log Likelihood		-143.8490	
AIC (smaller is better)		297.6980	
AICC (smaller is better)		301.4480	
BIC (smaller is better)		303.1533	

Algorithm converged.

Analysis Of Maximum Likelihood Parameter Estimates							
Parameter	DF	Estimate	Standard Error	Wald 95% Confidence Limits		Wald Chi-Square	Pr > ChiSq
Intercept	1	3.3610	0.6236	2.1388	4.5832	29.05	<.0001
sex	1	0.2814	0.3681	-0.4400	1.0028	0.58	0.4445
treatment	1	-0.3183	0.3261	-0.9575	0.3209	0.95	0.3291
baseline_n	1	0.0089	0.0013	0.0063	0.0115	43.64	<.0001
age	1	-0.0264	0.0243	-0.0741	0.0213	1.18	0.2776
Scale	0	3.3142	0.0000	3.3142	3.3142		

Note: The scale parameter was estimated by the square root of DEVIANCE/DOF.

The results are shown in Output 8.5. Comparing these with the results in Output 8.3, we see that the estimated regression coefficients are the same, but their standard errors are now much greater, with the consequence that only the coefficient of the baseline polyps count remains significant. Gender, treatment, and age are no longer found to be significant predictors of 3-month polyp count.

For a more detailed discussion of overdispersion, see Collett (2003).

8.5 Summary

Generalized linear models provide a very powerful and flexible framework for the application of regression models to medical data. Some familiarity with the basis of such models may allow medical researchers to consider more realistic models for their data rather than relying solely on linear and logistic regression.

8.6 Exercises

Exercise 8.1: Coronary Heart Disease Data

The data in the **chd** data set arise from a prospective study of potential risk factors for coronary heart disease (CHD) (Rosenman et al., 1975). The study looked at 3,154 men aged 40-50 for an average of 8 years and recorded the incidence of cases of CHD. The potential risk factors included smoking, blood pressure, and personality/behavior type.

Person-Years	Smoking	Blood Pressure	Behaviour	n of CHD Cases
5268.2	0	0	0	20

Person-Years	Smoking	Blood Pressure	Behaviour	n of CHD Cases
2542.0	10	0	0	16
1140.7	20	0	0	13
614.6	30	0	0	3
4451.1	0	0	1	41
2243.5	10	0	1	24
1153.6	20	0	1	27
925.0	30	0	1	17
1366.8	0	1	0	8
497.0	10	1	0	9
238.1	20	1	0	3
146.3	30	1	0	7
1251.9	0	1	1	29
640.0	10	1	1	21
374.5	20	1	1	7
338.2	30	1	1	12

Smoking: 0=non-smoker, 10=1-10 cigarettes a day, 20=11-20 cigarettes a day, 30=30+ cigarettes a day

Blood pressure: 0=<140, 1=$\geq$ 140

Behaviour: 0=Type B personality,1=Type A personality (Type A are characterized by impatience, competitiveness, aggressiveness, a sense of time urgency, and tenseness; Type B are characterized as being easy-going, relaxed about time, not competitive, and not easily angered or agitated.)

Apply a suitable Poisson regression model to these data. (Hint: Take another look at the analysis of the bladder cancer data in the text.) Summarize your conclusions as you would for a medical researcher who had collected the data.

Exercise 8.2: Gamma Distribution Model

Some of the counts in the polyp data in Table 8.1 are extremely large, indicating that the distribution of counts is very skewed. Consequently, the data might be better modelled by allowing for this with the use of a *gamma distribution* (the distribution is defined in Everitt and Skrondal, 2010). Apply such a model and compare the results with those given in the text derived from fitting a Poisson regression model. (Hint: Because gamma variables are positive, a log link function will again be needed.)

Exercise 8.3: Nested Models

Experiment with fitting different pairs of nested models to the FAP data and comparing the deviance values of the fitted models.

References

Agresti, A. *Introduction to Categorical Data Analysis*. New York: Wiley, 1996.

Aitkin, M. "The Analysis of Unbalanced Cross-Classifications." *Journal of the Royal Statistical Society. Series A* 142, no. 2(1978): 195–223.

Altman, D.G. *Practical Statistics for Medical Research*. London: Chapman & Hall/CRC, 1991.

Bruce, R.A., Kusumi, F. and Hosmer, D. "Maximal Oxygen Intake and Nomographic Assessment of Functional Aerobic Impairment in Cardiovascular Disease." *American Heart Journal* 84, no. 4(1973): 546–62.

Cleveland, W.S. *The Elements of Graphing Data*. Rev. ed. Murray Hill, NJ: Hobart Press, 1994.

Collett, D. *Modelling Binary Data*, 2nd Edition. London: Chapman & Hall/CRC, 2003.

Cook, R.D. and S. Weisberg. *Residuals and Influence in Regression*. New York: Chapman and Hall, 1982.

Der, G., and Everitt, B.S. *Applied Medical Statistics Using SAS*. Boca Raton, FL: CRC Press, 2013.

Dobson, A.J and Barnett, A.G. *An Introduction to Generalized Linear Models*, 3rd Edition. Boca Raton, FL: CRC Press, 2008.

Everitt, B.S. *The Analysis of Contingency Tables*, 2nd Edition. Boca Raton, FL: Chapman & Hall/CRC, 1992.

Everitt, B.S. *Statistical Methods for Medical Investigations*, 2nd Edition. London: E. Arnold, 1994.

Everitt, B.S. *Making Sense of Statistics in Psychology: A Second-Level Course*. Oxford: Oxford University Press, 1996.

220

Everitt, B.S. and Palmer, C.R. (Eds.) *The Encyclopaedic Companion to Medical Statistics*. London: Hodder Arnold, 2005.

Everitt, B.S. and Skrondal, A. *The Cambridge Dictionary of Statistics*. Cambridge: Cambridge University Press, 2010.

Fisher, R.A. *Statistical Methods for Research Workers*. Edinburgh; London: Oliver and Boyd, 1925.

Fleiss, J.L. *The Design and Analysis of Clinical Experiments*. New York, Wiley, 1999.

Freedman, W.L., Madore, B.F., Gibson, B.K. Ferrarese, L., Kelson, D.D., Sakai, S., Mould, J.R., Kennicutt, R.C., Ford, H.C., Graham, J.A., Huchra, J.P., Hughes , S.M.G., Illingworth, G.D., Macri, L.M. and Stetson, P.B. "Final Results from the Hubble Space Telescope Key Project to Measure the Hubble Constant." *The Astrophysical Journal* 533, no. 1(2001): 47–72.

Goldberg, D.P. *The Detection of Psychiatric Illness by Questionnaire: a Technique for the Identification and Assessment of Non-Psychotic Psychiatric Illness*. London: Oxford University Press, 1972.

Greenwood, M. and Yule, G.U. "An Inquiry into the Nature of Frequency Distributions Representative of Multiple Happenings with Particular Reference to the Occurrence of Multiple Attacks of Disease or of Repeated Accidents." *Journal of the Royal Statistical Society* 83, no. 2(1920): 255–79.

Hand, D.J., Daly, F., Lunn, A.D., McConway, K.J. and Ostrowski, E. (Eds.) *A Handbook of Small Data Sets*. London: Chapman and Hall, 1994.

Haberman, S.J. "The Analysis of Residuals in Cross-Classified Tables." *Biometrics* 29, no. 1(1973): 205–20.

Hosmer, D.W. and Lemeshow, S. *Applied Logistic Regression*, 2nd Edition. New York: Wiley, 2000.

Howell, D.C. *Statistical Methods for Psychology*, 5th Edition. Pacific Grove, CA: Duxbury/Thomson Learning, 2002.

Hurvich, C.M. and Tsai, C.L. "Regression and Time Series Model Selection in Small Samples."*Biometrika* 76, no. 2(1989): 297–307.

Kapoor, M. "Efficiency on Ergocycle in Relation to Knee-Joint Angle and Drag." Unpublished Master's Dissertation, University of Delhi (1981).

Kelsey, J.L. and Hardy, R.J. "Driving of Motor Vehicles as a Risk Factor for Acute Herniated Lumbar Intervertebral Disc." *American Journal of Epidemiology* 102, no. 1(1975): 63-73.

Liu, Z. "Smoking and Lung Cancer in China: Combined Analysis of Eight Case-Control Studies." *International Journal of Epidemiology* 21, no. 2(1992): 197–201.

McCullagh, P. and Nelder, J.A. *Generalized Linear Models*, 2nd Edition. London: Chapman and Hall, 1989.

McKay, R.J. and Campbell, N.A. "Variable Selection Techniques in Discriminant Analysis: I. Description." *British Journal of Mathematical and Statistical Psychology* 35, no. 1(1982): 1–29.

McKay, R.J. and Campbell, N.A. "Variable Selection Techniques in Discriminant Analysis: II. Allocation." *British Journal of Mathematical and Statistical Psychology* 35, no. 1(1982): 30-41.

Mann, L. "The Baiting Crowd in Episodes of Threatened Suicide." *Journal of Personality and Social Psychology* 41, no. 4(1981): 703–9.

Maxwell, S.E. and H.D. Delaney. *Designing Experiments and Analyzing Data*. Belmont, CA: Wadsworth, 1990.

Mehta, C.R. and N.R. Patel. "A Hybrid Algorithm for Fisher's Exact Test in Unordered r × c Contingency Tables." *Communications in Statistics – Theory and Methods* 15, no. 2(1986): 387–403.

Miles, J. and Shevlin, M. *Applying Regression and Correlation: A Guide for Students and Researchers.* London: Sage Publications, 2001.

Moore, D.S. "Statistics Among the Liberal Arts." *Journal of the American Statistical Association* 93, no. 444(1998): 1253-1259.

Moore, D.S. and McCabe, G.P. *Introduction to the Practice of Statistics.* New York: W.H. Freeman, 1989.

Nelder, J.A. "A Reformulation of Linear Models." *Journal of the Royal Statistical Society. Series A,* 140, no. 4(1977), 48-77.

Nelder, J.A. and Wedderburn, R.W.M. "Generalized Linear Models." *Journal of the Royal Statistical Society, Series A*, 135 no. 3(1972): 370-384.

Peterson, D.R, van Belle, G. and Chinn, N.M. "Epidemiologic Comparisons of the Sudden Infant Death Syndrome with Other Major Components of Infant Mortality." *American Journal of Epidemiology* 110, no. 6(1979): 699–707.

Piantadosi, S. *Clinical Trials: A Methodologic Perspective.* New York: Wiley, 1997.

Pine, R.W., Wertz, M.J., Lennard, E.S., Dellinger, E.P., Carrico, C.J. and Minshew, B.H. "Determinants of Organ Malfunction or Death in Patients with Intra-Abdominal Sepsis: A Discriminant Analysis." *Archives of Surgery* 118, no. 2(1983): 242–49.

Pugh, M.D. "Contributory Fault and Rape Convictions: Loglinear Models for Blaming the Victim." *Social Psychology Quarterly* 46, no. 3(1983): 233–42.

Rawlings, J.O., Pantula, S.G. and Dickey, D.A. *Applied Regression Analysis: A Research Tool.* New York: Springer, 2001.

Rickman, F., Mitchell, N., and Dingman, J. and Dalen, J.E. "Changes in Serum Cholesterol during the Stillman Diet." *The Journal of the American Medical Association* 228, no. 1(1974): 54–58.

Rosenman, R.H., Brand, R.J., Jenkins, C.D., Friedman, M., Straus, R. and Wurm, M. "Coronary Heart Disease in the Western Collaborative Group Study: Final Follow-up Experience of 8 ½ Years." *The Journal of the American Medical Association* 233, no. 8(1975): 872–77.

Scheffe, H. "A Method for Judging All Contrasts in the Analysis of Variance." *Biometrika* 40, no. 1-2(1953): 87–110.

Seeber, G.U.H. "On the Regression Analysis of Tumour Recurrence Rates." *Statistics in Medicine* 8, no. 11(1989): 1363–69.

Senie, R.T., Rosen, P.P., Lesser, M.L. and Kinne, D.W. "Breast Self-Examination and Medical Examination Related to Breast Cancer Stage." *American Journal of Public Health* 71, no. 6(1981): 583–90.

Tufte, E.R. *The Visual Display of Quantitative Information.* Cheshire, CT: Graphics Press, 1983.

van Belle, G., Fisher, L.D., Heagerty, P.J. and Lumley, T. *Biostatistics: A Methodology for the Health Sciences.* Hoboken, NJ: Wiley, 2004.

Venables, W.N. and Ripley, B.D. *Modern Applied Statistics with S-Plus.* New York: Springer-Verlag, 1994.

Wetherill, G.B. *Elementary Statistical Methods*, 3rd Edition. London: Chapman and Hall, 1982.

Woodley, W.L., Simpson, J., Biondini, R. and Berkeley, J. "Rainfall Results, 1970-1975: Florida Area Cumulus Experiment." Science 195, no. 4280(1977): 735–42.

Index

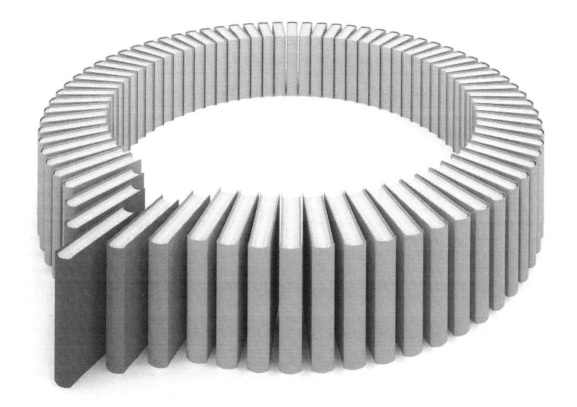

Gain Greater Insight into Your SAS® Software with SAS Books.

Discover all that you need on your journey to knowledge and empowerment.

support.sas.com/bookstore
for additional books and resources.

SAS and all other SAS Institute Inc. product or service names are registered trademarks or trademarks of SAS Institute Inc. in the USA and other countries. ® indicates USA registration. Other brand and product names are trademarks of their respective companies. © 2013 SAS Institute Inc. All rights reserved. S107969US.0413